bhbp

Carbohydrate Bioengineering
Interdisciplinary Approaches

In Memory of a Colleague and a Friend

by

Bernard Henrissat and Edward A. Bayer

On August 3, 2001 at about 1 am, our dear friend and colleague, Martin Schülein, suffered a massive heart attack and passed away quietly in his sleep.

Over the past two decades, Martin has been a unique international figure in the field of cellulases, and has served as a very special link between the research scientist and industry. He is a tremendous and irreplaceable loss to both.

We will remember Martin for his honesty, integrity and dignity. He demanded these not only of himself, but also of others, serving as a true role model for his peers. Because his opinion – be it positive or negative – was always profound and precise, it was always sought after and highly respected. Not a proponent of 'political correctness', his silence was as heavy as his statements. His entire manner radiated his innate enthusiasm and uncompromising sense of responsibility to his co-workers.

Martin was extremely faithful – not only to his scientific colleagues and friends – but also to his company, Novozymes (previously Novo/Nordisk). In his own strong but quiet way, he was very proud to be a Dane.

Carbohydrate Bioengineering
Interdisciplinary Approaches

Edited by

Tuula T. Teeri
Royal Institute of Technology, Stockholm, Sweden

Birte Svensson
Carlsberg Laboratory, Copenhagen, Denmark

Harry J. Gilbert
University of Newcastle upon Tyne

Ten Feizi
Imperial College, Harrow, UK

ROYAL SOCIETY OF CHEMISTRY

The proceedings of the 4th Carbohydrate Bioengineering Meeting held on the 10–13 June 2001 at the Royal Institute of Technology, Stockholm, Sweden.

Special Publication No. 275

ISBN 0-85404-826-X

A catalogue record for this book is available from the British Library

Published by The Royal Society of Chemistry,
Thomas Graham House, Science Park, Milton Road,
Cambridge CB4 0WF, UK
Registered Charity No. 207890

For further information see our web site at www.rsc.org

Printed by MPG Books Ltd, Bodmin, Cornwall, UK

Preface

Carbohydrates play a key part in essentially all living organisms where they can act as structural components of cell walls and reserves of stored energy as well as mediators of cell–cell interactions or the delicate control of protein folding among many other functions. Glycobiology and carbohydrate-active enzymes have wide-ranging applications in industry and medicine. Rapid technological development in glycobiology, molecular biology and, recently, genome sequencing and analysis have facilitated the design and manufacture of tailor-made carbohydrate molecules for the needs of, for example, the food, textile or pharmaceutical industries. The Carbohydrate Bioengineering Meetings, which are held biannually, provide regular up-dates into the current knowledge and the future directions of carbohydrate bioengineering. The previous three Carbohydrate Bioengineering Conferences, held in Elsinore, Denmark in 1995, LaRochelle, France in 1997 and Newcastle upon Tyne, UK in 1999, were very successful and attracted a substantial amount of industrial interest. The 4th Carbohydrate Bioengineering Meeting was organised in Stockholm, Sweden, June 10–13, 2001.

The Stockholm meeting attracted over 250 delegates mainly from Europe, Asia and North America. In addition to 37 lectures, both invited and selected based on abstracts, there were over 120 posters. This book contains a selection of papers based on the oral presentations. The first section is the keynote lecture addressing recent progress in the field of 'glycosynthases'. Owing to detailed knowledge of the structure and function of the active sites of given glycosyl hydrolases, it is possible to use protein engineering to direct their catalytic action towards glycan synthesis instead of hydrolysis. These new enzymes are now emerging as useful tools for tailor-made oligosaccharide synthesis for research and medicine.

The papers in Section 2 focus on newly determined three-dimensional structures of carbohydrate active enzymes and the subsequent functional studies. In addition to an increasing number of structures of different glycosyl hydrolases, structures of glycosyl transferases are now beginning to emerge. Section 3 extends to protein engineering of starch degrading enzymes, now possible owing to preceding detailed structure–function studies. Section 4 focuses on the study and engineering of the modular structures commonly found in various carbohydrate-active enzymes. Section 5 describes approaches for chemo-enzymatic oligosaccharide processing and synthesis.

The papers in the last two sections represent new trends in carbohydrate bioengineering. Section 6 discusses the enzymology of plant cell wall carbohydrates. Section 7 touches upon trends in the field of glycobiology emerging from work on the genome and the glycome.

The meeting organisers gratefully acknowledge the financial support from the following organisations: *Bo Rydin Foundation for Scientific Research, Swedish Natural Science Research Council, Swedish Council for Forestry and Agricultural Research, Royal Institute of Technology (KTH), Applied Biosystems, Genencor International Inc., Neose Technologies Inc., Micromass AB, Novozymes, Eka Chemicals, Ezaki Glio, Co, Ltd, Amano Enzyme Inc., Nihon Shokuhin Kako Co, Ltd, Nagase Biochemicals Ltd* and *Ozeki Corporation*. We are indebted to Stuart Denman for proficient maintenance of the registration services and abstract databases among many, many other things – always with a good joke at hand! We are grateful for Mrs Alphonsa Lourdudoss for valuable help with the economic management of the meeting. The KTH Conference Service is thanked for professional help with the website and the practical arrangements. The staff and students in the KTH Wood Biotechnology Laboratory are warmly thanked for efficient help – with a smile – upon registration, at the conference office and during the breaks. Finally, we wish to thank all speakers, authors and poster presenters for their contributions as well as the delegates for attending the meeting. We hope to see you all again in the Netherlands in 2003!

Tuula Teeri, Birte Svensson, Harry Gilbert and Ten Feizi; the Editors.

Contents

4 Domain Structure and Engineering

5 Chemo-enzymatic Carbohydrate Synthesis

1 Keynote Address

ENGINEERING GLYCOSIDASES FOR CONSTRUCTIVE PURPOSES

David L. Jakeman and Stephen G. Withers

Department of Chemistry, Protein Engineering Network of Centres of Excellence, University of British Columbia, Vancouver, Canada.

1 ABSTRACT

The synthesis of oligosaccharides using glycosidases as catalysts has been known for many years, but has been limited by the poor yields generally associated with such processes due to hydrolysis. Glycosynthases, mutant glycosidases in which the catalytic nucleophile has been replaced, offer an alternative form of catalyst that synthesizes glycosides from readily prepared glycosyl fluorides, but does not hydrolyze the glycoside products. The range of reactions performed by these enzymes is being extended through mutation of number of different glycosidases, as well as through random mutation of known glycosynthases, coupled with efficient screens.

2 DISCUSSION

Glycosidases have important roles as industrial catalysts for the breakdown of complex carbohydrates, yet their commercialization as catalysts for the synthesis of oligosaccharides has been less fruitful because the product is necessarily a substrate for hydrolysis by the wild-type enzyme. Despite the general lack of commercial success in exploiting glycosidases as catalysts for the synthesis of oligosaccharides, a diverse selection of glycosidases has been explored in the literature to perform transglycosylation reactions yielding oligosaccharides, and this area of research has been reviewed recently.[1] There are many possible parameter permutations available to increase the yields of glycosidase-mediated transglycosylation, and for specific glycosidases respectable yields have been observed as a result of altering certain reaction conditions, but no one process has been universally successful in circumventing the hydrolysis conundrum and giving consistently high yields.

The mechanism of retaining glycosidases has been the study of intense research[2] initially based upon Koshland's premise of a double-displacement occurring at the anomeric centre.[3] Figure 1A shows the widely accepted reaction course for a retaining glycosidase hydrolyzing a substrate. Two active site carboxylic acid residues separated by approximately 5.5 Å apart act as a nucleophile and general acid / base catalyst respectively. Hydrolysis is initiated by attack of the nucleophile onto the anomeric carbon to form a covalent glycosyl-enzyme intermediate that in turn is hydrolyzed by an incoming water molecule to release enzyme and monosaccharide. The transglycosylation mechanism differs from the hydrolytic mechanism solely because a sugar residue replaces the water molecule attacking the covalent glycosyl enzyme-intermediate (Figure 1B). One approach taken to avoid hydrolysis of product from a transglycosylation reaction was the abolition of the first step of the hydrolytic reaction - by mutating the catalytic nucleophile to a non-nucleophilic residue.[4] Whilst this approach does not

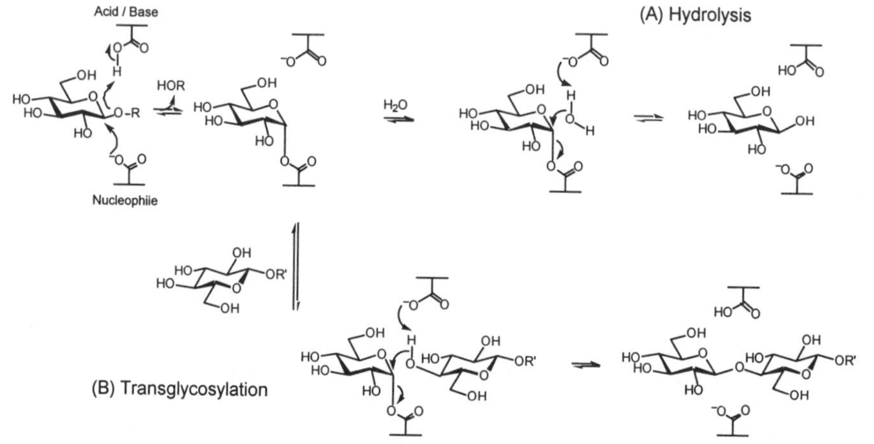

Figure 1 *(A) Hydrolysis and (B) transglycosylation by a retaining glycosidase*

permit formation of a covalent glycosyl-enzyme intermediate, an active site architecture is maintained that could condense an appropriate donor and acceptor sugar together provided the donor mimics the glycosyl enzyme-intermediate. Because the placement of the acid / base catalyst within the active site has not been altered by the mutation, it remains set up to deprotonate an incoming acceptor sugar just as it does in a transglycosylation reaction (Figure 1B). The discovery that glycosyl fluorides function as suitable donors for reactions catalyzed by specific nucleophile mutants of retaining glycosidases has now resulted in several new mutant enzymes capable of catalyzing the synthesis of oligosaccharides.[4-10] A proposed mechanism for glycosynthases is shown in Figure 2, as postulated for the first glycosynthase *Agrobacterium* sp. β-glucosidase (Abg) E358A.[4] The donor sugar composed of α-D-glucosyl fluoride, of equivalent anomeric configuration to the intermediate, binds in the glycone-binding site mimicking the intermediate. The acid / base residue facilitates attack of the nucleophilic hydroxyl group onto the anomeric carbon atom by deprotonating the acceptor sugar. Fluoride is released, and a glycosidic linkage is formed.

The catalytic promiscuity of Abg E358A was initially uncovered by the discovery that it will transfer α-D-glucosyl fluoride (GlcF) and α-D-galactosyl fluoride (GalF) onto a variety of acceptor sugars in moderate to high yields. Such is the efficiency of the carbohydrate transfer that with GlcF as donor repeated additions occur, providing a series of cello-oligosaccharides. Of the compounds generated utilizing Abg E358A as catalyst,

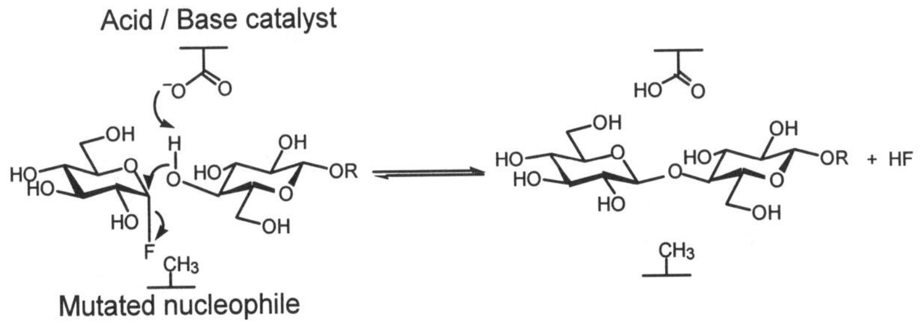

Figure 2 *Proposed glycosynthase mechanism catalyzed by Abg E358A. R = 2-nitrophenyl*

arguably the most noteworthy are mechanism-based inactivator for cellulases, 2,4-dinitrophenyl 2-deoxy-2-fluoro-β-D-cello-oligosides (Figure 3). These compounds would not be amenable to synthesis using a traditional wild-type transglycosylation approach because the 2-deoxy-2-fluoro glycoside rapidly inactivates the wild-type enzyme by covalent modification.[4]

Further advances in rates of glycosylation were achieved by replacement of the nucleophile by a serine residue.[6] Abg E358S was a 24-fold faster catalyst as determined by fluoride ion release and this rate enhancement

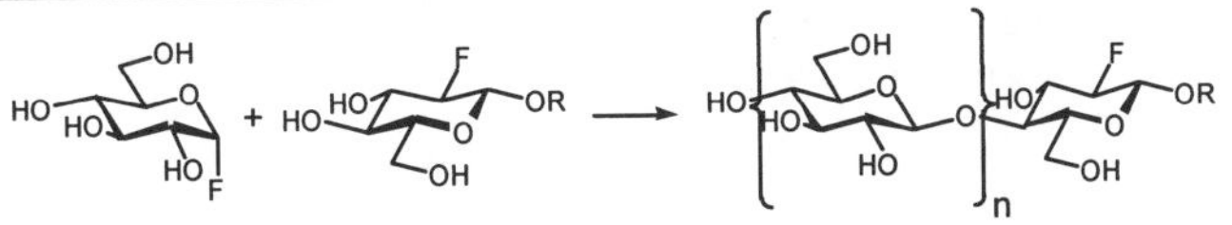

Figure 3 *Abg E358A catalyst, 84 % yield (n = 1, 38 %; n = 2, 42 %; n = 3, 4 %). R = 2,4-dinitrophenyl*

was attributed to stabilization of the departing fluoride *via* hydrogen bonding. Importantly, the increase in transglycosylation rate also broadened the substrate specificity of the glycosynthase catalyst. The condensation of GalF and 4-NP β-*N*-

Figure 4 *Abg E358S catalyst. 63 % yield. R = 2-nitrophenyl*

acetylglucosaminide to give 4-NP β-*N*-acetyllactosaminide (Figure 4) was only successful using the E358S mutant enzyme.

We have developed a novel screen to select for improved glycosynthase activity and used it to successfully discover two further glycosynthase mutants of Abg.[5] The screen requires an *endo*-cellulase to selectively hydrolyze products derived from a glycosynthase reaction. We chose Cel5A from *Cellulomonas fimi*, (*C. fimi*) an enzyme that has exquisite substrate specificity for di- or trisaccharide substrates. Thus, the screen works through glycosynthase catalyzed condensation of GlcF and 4-methylumbelliferyl β-D-glucoside (MUGlc) to provide a disaccharide product that is then hydrolyzed by the cellulase releasing the fluorescent methylumbelliferone (Figure 5). In total four mutants were selected by screening a saturation mutagenesis library of Abg E358X (X = any amino acid) prepared by a four-primer PCR strategy. Two of these mutants were the known alanine and serine glycosynthases and two were new glycosynthases, the cysteine and glycine mutants. The E358C mutant was a very poor glycosynthase with a rate only half that of the E358A mutant, whilst the E358G mutant

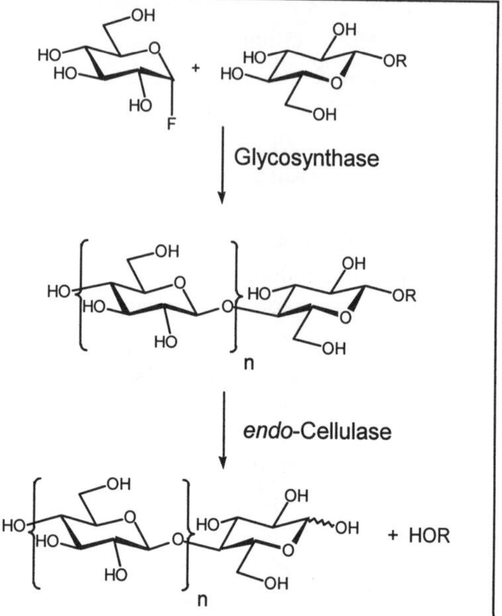

Figure 5 *A glycosynthase selection using Abg glycosynthase and endo-cellulase Cel5A. R = methylumbelliferyl*

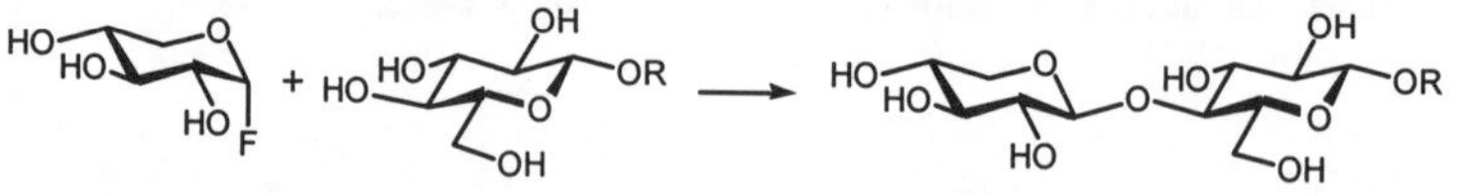

had a rate twice that of the E358S mutant. The increase in transglycosylation rate was sufficient to permit the condensation of a new glycosyl fluoride by Abg E358G. Thus, α-D-xylosyl fluoride was condensed with 4-nitrophenyl glucoside in 43 % isolated yield (Figure 6). While the yield for the reaction is not as impressive as with many of the glycosynthase catalyzed reactions, the discovery of a new glycosyl fluoride donor is particularly significant because the mutant was selected by screening for a different

Figure 6 *Abg E358G catalyzes transfer of a new glycosyl donor, xylosyl fluoride. 43 % yield. R = 2-nitrophenyl*

specific reaction. Applying *in vitro* evolution techniques to the whole Abg gene will hopefully provide new and improved glycosynthase catalysts with transferase behavior corresponding to the broad substrate specificity afforded by wild-type Abg.[11]

The β-mannosidic linkage is one of the most chemically challenging glycosidic linkages to obtain with high anomeric stereoselectivity. The development of a glycosynthase to efficiently condense α-D-mannosyl fluoride (ManF) with numerous acceptor sugars[8] has provided a new efficient route to complement the chemical strategies available in the literature. The mannosidase from *C. fimi* Man2a was cloned and expressed in *E. coli* and the nucelophile identified as E519.[12] Two active site mutants were prepared, E519A and E519S, and both mutants exhibited glycosynthase behavior. However, yields were significantly higher with the serine mutant.[8] Essentially quantitative conversion to oligosaccharide product was observed when ManF and 4-nitrophenyl β-D-cellobioside were condensed (Figure 7) providing ready access to a new class of oligosaccharides. Other aryl glycosides were also suitable acceptors for Man2a E519S, including 4-nitrophenyl β-D-mannoside, 4-nitrophenyl β-D-xyloside and 4-nitrophenyl β-D-glucoside. Overall yields were high for oligomeric products and when 2,4-dinitrophenyl 2-deoxy-2-fluoro-β-D-mannoside was used as acceptor a hexameric oligosaccharide species was isolated resulting from the transfer of five ManF units. Often the transfer of the second ManF sugar was achieved efficiently but with loss of absolute regiochemical control because -β(1,3)- and -β(1,4)- linkages were generated. One reaction that did not produce a mixture of regiochemical isomers was the condensation of ManF and 4-nitrophenyl β-D-gentiobioside (Figure 8). High yields of oligomeric products were observed providing a route into novel branched oligosaccharides.

In the literature, glycosynthases have been developed that transfer a number of different donor sugars. The first reported glycosynthase derived from an *endo*-glycosidase was the retaining 1,3-1,4-β-glucanase from *Bacillus licheniformis* E134A. The mutant enzyme catalyzed the condensation of α-laminaribiosyl fluoride onto MUGlc in 88 % yield.[10]

The glycosynthase from *Humicola insolens* Cel7B E197A is a particularly well characterized glycosynthase,[9] capable of synthesizing a wide range of oligo- and polysaccharides from several different donor sugars and remains the only glycosynthase for which an X-ray crystal structure has been reported. It condenses α-cellobiosyl fluoride with great efficiency into cellulose II. Defined oligosaccharide products have also been obtained when condensing α-lactosyl fluoride (LacF) onto a range of disaccharide and monosaccharide acceptor sugars. Quantitative yields were obtained using methyl β-D-cellobioside, methyl 6-bromo-6-deoxy-cellobioside and benzyl laminaribioside as acceptors while lower yields were obtained using monosaccharide acceptors. From the analysis of the wild-type crystal structure it was observed that specific hydroxyl groups on the substrate molecules failed to interact significantly with the polypeptide chain. These relaxed constraints in the substrate permitted the condensation of novel α-glycosyl fluorides modified at the 6^I and 6^{II} position to afford novel polymers.

3 CONCLUSIONS

The mutation of a retaining glycosidase to a hydrolytically inactive enzyme by substituting the nucleophilic carboxylic acid with a non-nucleophilic residue has generated a novel class of glycosylation catalysts. These mutant enzymes transfer activated donor sugars onto a wide variety of acceptor sugars with high degrees of stereo- and regioselective control and in very good yield. The synthesis of donor sugars is straightforward, using cheap and readily available starting materials. Glycosynthases, therefore, provide a new route for the large-scale manufacture of oligosaccharides. A wide variety of oligosaccharides can easily be prepared using glycosynthase technology.

However, because of the large amount of sequence information on glycosidases,[13] it is surely only a matter of time before new glycosynthases are reported, further broadening the range of donor sugars that can be transferred and providing a means to synthesize other more exotic glycosidic linkages.

4   ACKNOWLEDGEMENTS

We thank the Protein Engineering Network of Centres of Excellence of Canada, the Natural Sciences and Engineering Research Council of Canada and Neose Technologies Inc. for financial support.

References

1 Vocadlo, D. J., & Withers, S. G. (2000) in *Carbohydrates in Chemistry and Biology* (Ernst, B., Hart, G. W., & Sinay, P., Eds.) pp 723, Wiley-VCH GmbH, Weinheim, Germany.

2 Zechel, D. L., & Withers, S. G., *Acc. Chem. Res.,* 2000, **33**, 11 .

3 Koshland, D. E., *Biol. Rev.,* 1953, **28**, 416.

4 Mackenzie, L. F., Wang, Q. P., Warren, R. A. J., & Withers, S. G., *J. Am. Chem. Soc.,* 1998, **120**, 5583 .

5 Mayer, C., Jakeman, D. L., Mah, M., Karjala, G., Gal, L., Warren, R. A. J., & Withers, S. G., *Chem. Biol.,* 2001, In the press.

6 Mayer, C., Zechel, D. L., Reid, S. P., Warren, R. A. J., & Withers, S. G., *FEBS Lett.,* 2000, **466**, 40 .

7 Trincone, A., Perugino, G., Rossi, M., & Moracci, M., *Bioorg. Med. Chem. Lett.,* 2000, **10**, 365 .

8 Nashiru, O., Zechel, D. L., Stoll, D., Mohammadzadeh, T., Warren, R. A. J., & Withers, S. G., *Angew. Chem. Int. Ed.,* 2001, **40**, 417 .

9 Fort, S., Boyer, V., Greffe, L., Davies, G., Moroz, O., Christiansen, L., Schulein, M., Cottaz, S., & Driguez, H., *J. Am. Chem. Soc.,* 2000, **122**, 5429 .

10 Malet, C., & Planas, A., *FEBS Lett.,* 1998, **440**, 208 .

11 Kempton, J. B., & Withers, S. G., *Biochemistry,* 1992, **31**, 9961 .

12 Stoll, D., He, S., Withers, S. G., & Warren, R. A., *Biochem. J.,* 2000, **351 Pt 3**, 833 .

13 , http://afmb.cnrs-mrs.fr/~pedro/CAZY/.

2 Structure–Function Studies of Carbohydrate-active Enzymes

STRUCTURAL ENZYMOLOGY OF CARBOHYDRATE-ACTIVE ENZYMES

Gideon J. Davies

Structural Biology Laboratory
Department of Chemistry
University of York
York, YO10 5DD
U.K.

1 INTRODUCTION

Simply in terms of quantity, the enzymatic synthesis and degradation of glycosidic bonds are the most important reactions on earth. The enzymes involved both in these processes and in the modification of oligo- and polysaccharides - "carbohydrate-active enzymes" or CAZymes - have thus received long-standing interest from both academic and industrial groups. Glycoside hydrolases, the degradative machinery have recently been extensively reviewed[1-4] and will not be covered here. Instead I discuss the structures and functions of some other enzyme classes involved in the synthesis and degradation of poly and oligosaccharides: polysaccharide lyases, carbohydrate esterases and, natures synthetic apparatus, glycosyltransferases. Before embarking on the structures and likely mechanisms, however, it is worth considering both the amino-acid sequence families and modularity of CAZymes which underpins all work in this field.

One of the most important breakthroughs in our understanding of CAZymes was the classification system, based upon amino-acid sequence similarities, proposed by Bernard Henrissat in the late 1980's[5]. Originally, this was a classification of glycoside hydrolases only, based upon early work with cellulases,[6,7] but it was later extended to include glycosyltransferases[8], the many ancillary modules, particularly those involved in carbohydrate binding (described below)[9] as well as other enzyme classes. This continually-updated resource is available at URL http://afmb.cnrs-mrs.fr/~pedro/CAZY/db.html and currently provides classification of over 86 families of glycoside hydrolases, 54 glycosyltransferases, 12 polysaccharide lyases, 13 families of carbohydrate-active esterases and (de)acetylases and almost 30 families of carbohydrate-binding modules. This site also provides information both on the catalytic mechanism of each family, where known and on the availability three-dimensional structures for the various modules.

Whilst the concept of modular proteins appears only to be dawning slowly in mainstream biochemistry,[10] it is to the credit of the glycoside hydrolase community that both the prevalence and importance of modularity was recognised long ago, as has been recently been reviewed.[11] It soon became clear that many enzymes were multi-modular proteins, in which a catalytic module was fused to one or more ancillary domains. The function of many of these ancillary modules was shown to be carbohydrate-binding and has, thus far, led to the classification of 29 sequence-based families of carbohydrate-binding modules (CBMs).[9] There also remain almost 70 identified families of "X" modules whose function is, as yet, unknown. Understanding this modularity clearly has enormous implications in the post-genomic age if we are ever even to annotate open-reading frames correctly, let alone determine their function.[11]

Individual modules may either be linked by flexible linker sequences (with the result that 3-D structural study has proved rather challenging) or the individual modules may be more rigidly associated with each other: this latter situation having been revealed in the complex modular architecture of many glycoside hydrolases such as β-galatosidase,[12] chitobiase,[13] and cyclodextrin transglycosylases,[14] for example. Structures of intact multi-domain proteins in which the linker is flexible have rarely been achieved: a notable exception being the family 10 xylanase from *Streptomyces olivaceoviridis* with its intact xylan binding domain.[15]

In contrast to aerobic organisms, which secrete discrete multi-modular proteins into the extra-cellular environment, the plant cell-wall degradative apparatus of some anaerobic organisms are even more complicated. They may be organised in a supramolecular multi-protein complex termed the cellulosome.[16] The cellulosome is certainly elaborate: a variety of enzymes, which are themselves multi-modular, are attached *via* their "dockerin" domains to the "cohesin" domains of a macromolecular scaffold, Figure 1. This "scaffoldin" protein may in turn be tethered to the bacterial cell.

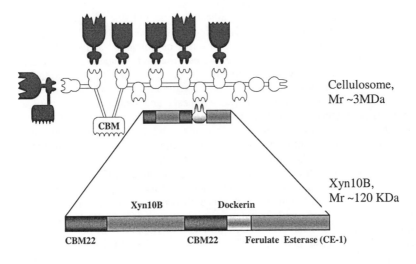

Cellulosome,
Mr ~3MDa

Xyn10B,
Mr ~120 KDa

Key to symbols

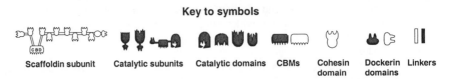

Scaffoldin subunit   Catalytic subunits   Catalytic domains   CBMs   Cohesin domain   Dockerin domains   Linkers

Figure 1 *The often complex modularity of carbohydrate-active enzymes. Original cellulosome figure kindly provided by Professor Ed Bayer.*

Together, this makes for a rather complex organisation for many CAZymes which must first be dissected if we are to understand the individual modules and their biochemical functions.[11]

2 THE ENZYMATIC SYNTHESIS OF GLYCOSIDIC BONDS

Enzymatic glycosyl transfer is, quantitatively, the most significant reaction on earth. It is catalysed by glycosyltransferases, which utilise activated glycosyl donors in order to drive the reaction in aqueous solution. The activating group may be phosphate, nucleoside-phosphate or lipid-phosphate. Both inverting enzymes, whose product displays the opposite stereochemistry at the anomeric centre to that of the activated donor, and retaining transferases, whose product retains the same anomeric configuration as the donor, are known. Currently 56 families of glycosyltransferase are described, yet at time of the last Carbohydrate Bioengineering Meeting in 1999 only two true glycosyltransferase structures had been reported: the phage T4 β-glucosyltransferase[17] and the SpsA protein from *Bacillus subtilis*[18]. Since then, structures for inverting enzymes from families GT-1,[19] family GT-7 (the bovine β-galactosyltransferase),[20] family GT-13 (the rabbit *N*-acetyl glucosaminyltransferase I)[21] the family GT-28 MurG enzyme[39] and family GT-43 human β–1,3 glucuronyltransferase I[22] have been reported. The first two retaining enzyme structures have also now been determined, those from family GT-8 (the Neisserial galactosyltransferase LgtC)[23,24] and family GT-6 (the bovine α-1,3-Galactosyltransferase).[25]

A striking feature of the inverting transferase families now determined is that they appear to fall into just two structural superfamilies exemplified by the original T4 Glc transferase and family GT-2 SpsA structures, Figure 2.

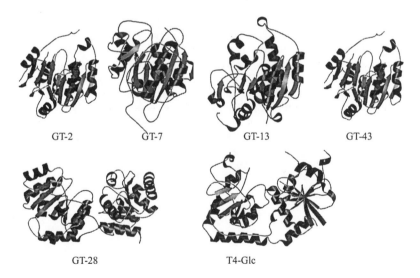

Figure 2 *3-D structures of glycosyltransferases from families 2, 7, 13, 28 & 43 as well as the phage T4 glycosyltransferase. Thus far inverting transferases appear to fall into just two superfamilies.*[26-29]

An example of the conserved catalytic machinery within a "superfamily or "clan" of glycosyltransferases is given below, Figure 3. Families 2,7,13 and 42 all appear to share the same catalytic apparatus and reaction mechanism[30].

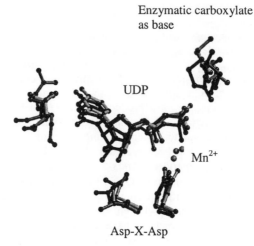

Figure 3 *Overlay of the conserved catalytic machinery seen in glycosyltransferase families 2, 7, 13 and 43.*

3 POLYSACCHARIDE LYASES

Polysaccharide lyases (E.C. 4.2.2.x) are carbon-oxygen lyases that cleave glycosidic bonds of C5-uronic acid containing pyranoside substrates and are widely believed to utilise β-elimination chemistry. They play a pivotal role in the biological recycling of plant material, a process essential for biosphere maintenance and are potent virulence factors of eukaryotic pathogens. Henrissat currently describes twelve families of lyases (http://afmb.cnrs-mrs.fr/~pedro/CAZY/lya.html). The essential enzymatic feature of lyases is that they have a requirement for sugar uronic acids: the carboxylate substituent being essential to allow "acidification" and hence abstraction of the proton at C5 and elimination of the leaving group, Figure 4.

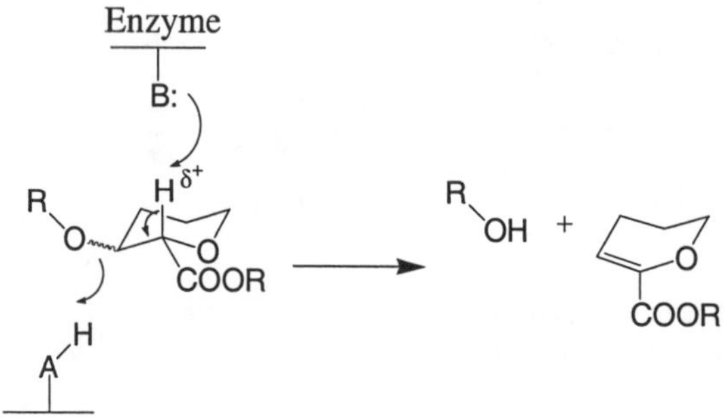

Figure 4 *Generic catalytic mechanism for a polysaccharide lyase.*

Three dimensional (3D) structures are available for PL 1, 3, 5, 6 and 8 enzymes and divide these families into those possessing either an "α/α", PL 5 and 8, or "parallel β-helix", PL 1, 3 and 6, fold. Despite this structural resource and accompanying biochemical analyses, there is as yet little data directly supporting the commonly hypothesised catalytic mechanism that features proton abstraction from C5 of the +1 subsite sugar residue, termed the α-carbon, and proton donation to the glycosidic oxygen, leading to the elimination of a second group from C4, termed the β-carbon[31]. Indeed, it is unclear whether there is negative charge build-up on the glycosidic oxygen at the transition-state and hence the requirement for proton donation by an acid catalyst, although much speculated about, is as yet undemonstrated.

We recently solved the structure of a novel pectate lyase from family PL-10.[32,33] The structure is a variant on the α/α barrel in marked contrast to the parallel β-helix fold displayed by all other previous pectate lyases. Despite this structural dissimilarity, the structure determination, of PL10 in complex with an unhydrolysed trisaccharide substrate, points to a structurally invariant arginine residue which is also conserved in the totally-unrelated pectate lyase structures from family 1. This may be indicative of a conserved catalytic mechanism acquired through convergent evolution and helps point the way towards useful mechanistic studies on this class of enzymes.

4 CARBOHYDRATE ESTERASES

The Henrissat classification includes thirteen different families of sugar esterases and de-*N* / de-*O* acetylases, although it is clear that other families may exist. The de-esterification or de-acetylation reaction is similar to the hydrolysis of lipids and peptides and it is therefore not surprising that the majority of the carbohydrate-active enzymes bare a great deal of similarity to known lipases and proteases. The most common catalytic mechanism described thus far is, therefore, the classical serine-protease-like mechanism based upon the Ser-His-Asp catalytic triad and the α/β hydrolase "lipase" fold. The hydrolysis reaction catalysed by such enzymes occurs *via* the formation and subsequent breakdown of a covalent acyl-enzyme intermediate (reviewed in[34,35]). Of particular importance in the plant-cell wall hydrolysis field is how the different enzymes display specificity for certain substrates. Some enzymes are multi-functional whereas others exhibit an extremely narrow substrate range. A major challenge, for example, has remained the elucidation of the structure of a ferulate esterase and the basis of its specificity for the different cinnamic acids found in the plant cell-wall.

Xylan is the most abundant and important hemicellulose. It is comprised of a backbone of β-1,4-linked D-xylopyranose units decorated, at O2 and/or O3, with L-arabinofuranose, acetate or 4-*O*-methyl glucuronic acid moieties. Linkage between xylan and lignin, two major plant cell wall components, is established by cinnamic acids, which bridge the arabinofuranosyl side-chains of the former to the latter. The ester bond between the arabinofuranoside and ferulic acid (3-methoxy, 4-hydroxy cinnamic acid) is hydrolysed by feruloyl esterases. These enzymes both aid the release of hemicellulose from lignin and also render the free polysaccharide product more amenable to degradation by other biocatalysts such as endo-β1,4-xylanases and arabinofuranosidases.

The recent structure determination of the *Clostridium thermocellum* Xyn10B ferulate esterase domain (shown in Figure 1)[36] revealed both the catalytic machinery and the determinants of substrate specificity for this enzyme, Figure 5. The *C. thermocellum* Xyn10A ferulate esterase has also, simultaneously, been solved by the Ljungdahl group in Georgia and together these two structures should soon help shed light on ferulate recognition and hydrolysis and the degradation of the plant cell wall.

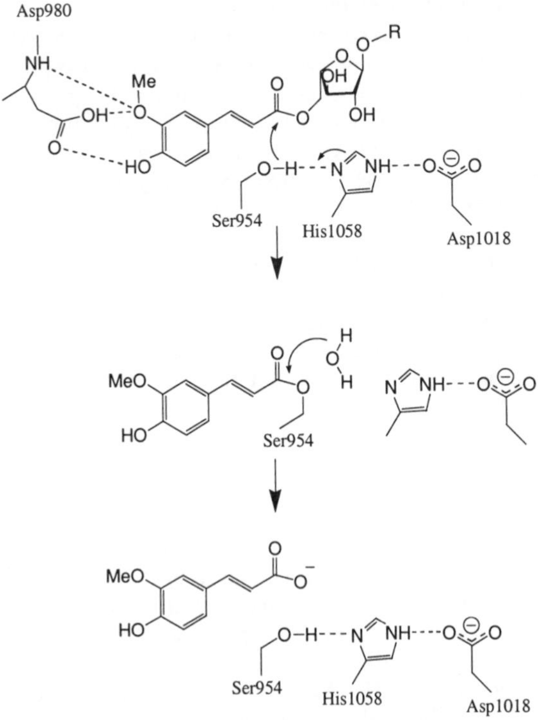

Figure 5 *The reaction catalysed by the C. thermocellum Xyn10B ferulate esterase module.*

5 SUMMARY

The world of carbohydrate-active enzymes remains a complex one. In the post genomic era we are inundated with new sequences and ORFs of unknown function. The modular nature of many of these enzymes provides a challenge both for biochemical characterisation and for functional genome annotation. The sequence-derived classifications of Henrissat and co-workers go a long way to answering some of these challenges. At the structural level, our understanding of hydrolases grows: especially now even hen-egg white lysosyme has been brought back into the fold and shown to share a common catalytic mechanism with other retaining hydrolases[37]! Many hydrolases remain unstudied or poorly understood. Indeed, whilst our knowledge of glycoside hydrolase catalytic mechanism remains superficially good, it is not yet sufficient to allow the routine synthesis of selective transition-state mimics as mechanistic probes and therapeutic agents – and many carbohydrate-active enzymes are major drug targets.[38] Our understanding of the other enzymes involved in the synthesis and degradation of oligo- and polysaccharides lags behind. We are just beginning to witness an increase in the number of 3-D structures for esterases, lyases, transferases and their respective carbohydrate-binding modules, but there is much to be done before the structural enzymology of carbohydrate-active enzymes is fully understood.

6 ACKNOWLEDGEMENTS

The author would like to thank Bernard Henrissat for useful discussions and both the Wellcome Trust and the BBSRC for supporting the Structural Biology Laboratory. G.J.D is a Royal Society University Research Fellow. The work reviewed here was performed in York by Simon Charnock, Valérie Ducros, Nicolas Tarbouriech, Johan Turkenburg and Annabelle Varrot. Much of the work is a collaboration with the groups of Gary Black (University of Northumbria), Carlos Fontes (Universitário do Alto da Ajuda, Lisbon), Harry Gilbert (University of Newcastle upon Tyne), Steve Withers (University of British Columbia, Vancouver) and the late and sadly missed Martin Schülein at Novozymes A/S. Such collaborative work would not have been possible without the provision of Structural Biology Centre funding from the BBSRC.

References

1. Davies, G., Sinnott, M. L. & Withers, S. G. in *Comprehensive Biological Catalysis* ed. M. L. Sinnott, Academic Press, London, 1997, pp. 119.
2. Rye, C. S. & Withers, S. G. *Curr. Op. Chem. Biol.* 2000 **4**, 573.
3. Zechel, D. & Withers, S. G. in *Comprehensive Natural Products Chemistry* eds. D. Barton, K. Nakanishi, & C.D. Poulter, Elsevier Science, Amsterdam, The Netherlands, 1999, pp. 279.
4. Zechel, D. L. & Withers, S. G. *Acc. Chem. Rev.* 2000 **33**, 11.
5. Henrissat, B. & Davies, G. J. *Curr. Op. Struct. Biol.* 1997 **7**, 637.
6. Henrissat, B. *Biochem. J.* 1991 **280**, 309.
7. Henrissat, B. & Bairoch, A. *Biochem. J.* 1993 **293**, 781.
8. Campbell, J. A., Davies, G. J., Bulone, V. & Henrissat, B. *Biochem. J.* 1997 **326**, 929.
9. Coutinho, P. M. & Henrissat, B. in *Recent Advances in Carbohydrate Engineering* eds. H.J. Gilbert, G. J., Davies, B. Svensson, B. & B. Henrissat, Royal Society of Chemistry, Cambridge, U.K., 1999, pp.3.
10. Khosla, C. & Harbury, P. B. *Nature* 2001 **409**, 247.
11. Henrissat, B. & Davies, G. J. *Plant Physiol.* 2000 **124**, 1515.
12. Jacobson, R. H., Zhang, X.-J., DuBose, R. F. & Matthews, B. W. *Nature* 1994 **369**, 761
13. Tews, I., Perrakis, A., Oppenheim, A., Dauter, Z., Wilson, K. S. & Vorgias, C. E. *Nat. Struct. Biol.* 1996 **3**, 638
14. Uitdehaag, J. C. M., Mosi, R., Kalk, K. H., van der Veen, B. A., Dijkhuizen, L., Withers, S. G. & Dijkstra, B. W. *Nat. Struct. Biol.* 1999 **6**, 432.
15. Fujimoto, Z., Kuno, A., Kaneko, S., Yoshida, S., Kobayashi, H., Kusakabe, I. & Mizuno, H. *J. Mol. Biol* 2000 **300**, 575.
16. Bayer, E. A., Chanzy, H. & Lamed, R. *Curr. Op. Struct. Biol.* 1996 **8**, 548.
17. Vrielink, A., Rüger, W., Driessen, H. P. C. & Freemont, P. S. *EMBO J* 1994 **13**, 3413.
18. Charnock, S. & Davies, G. J. *Biochemistry* 1999 **38**, 6380.
19. Mulichak A.M., Losey H.C., Walsh C.T. & R.M., Garavito. *Structure* 2001 **9**, 547.
20. Gastinel, L. N., Cambillau, C. & Bourne, Y. *EMBO J.* 1999 **18**, 3546.
21. Ünligil, U. M., Zhou, S., Yuwaraj, S., Sarkar, M., Schachter, H. & Rini, J. M. *EMBO J.* 2000 **19**, 5269.
22. Pedersen, L. C., Tsuchida, K., Kitagawa, H., Sugahara, K., Darden, T. A. & Negishi, M. *J. Biol. Chem.* 2000 **275**, 34580.

23. Persson, K., Ly, H. D., Dieckelmann, M., Wakarchuk, W. W., Withers, S. G. & Strynadka, N. C. J. *Nat. Struct. Biol.,* 2001 **8**, 166.
24. Davies, G. J. *Nat. Struct. Biol.,* 2001 **8**, 98.
25. Gastinel, L. N., Bignon, C., Misra, A. K., Hindsgaul, O., Shaper, J. H. & Joziasse, D. H. *EMBO J.* 2001 **20**, 638.
26. Ünligil, U. M. & Rini, J. M. *Curr. Op. Struct Biol.* 2000 **10**, 510.
27. Bourne, Y. & Henrissat, B. *Curr. Op. Struct Biol.* 2001 in press, October 2001.
28. Charnock, S. J., Henrissat, B. & Davies, G. J. *Plant Physiol.* 2001 **125**, 527.
29. Davies, G. J., Charnock, S. J. & Henrissat, B. *Trends Glycosci. Glycotechnol.* 2001 **13**, 1.
30. Tarbouriech, N., Charnock, S. J. & Davies, G. J. submitted to *J. Mol. Biol* 2001.
31. Anderson, V. E. in *Comprehensive biological catalysis* ed. Sinnott, M. Academic press, London, 1998, pp. 115.
32. Brown, I. E., Mallen, M. H., Charnock, S. J., Davies, G. J. & Black, G. W. *Biochem. J.* 2001 **355**, 155.
33. Charnock, S. J., Brown, I. E., Turkenburg, J. P., Black, G. W. & Davies, G. J. *Acta Crystallogr.* 2001 **D57**, 1141.
34. Dodson, G. & Wlodawer, A. *Trends Biochem. Sci.* 1998 **23**, 347.
35. Wharton, C. W. in *Comprehensive Biological Catalysis* ed. M.L. Sinnott, Academic Press, London, 1998, pp345.
36. Prates, J. A. M., Tarbouriech, N., Charnock, S. J., Fontes, C. M. G. A., Ferreira, L. M. A. & Davies, G. J. submitted to *Structure* 2001.
37. Vocadlo, D. J., Davies, G. J., Laine, R. & Withers, S. G. *Nature* 2001 **412**, 835.
38. Williams, S. J. & Davies, G. J. *Trends Biotechnol.* 2001 **19**, 356.
39. Ha, S., Walker, D., Shi, Y. & Walker, S. *Prot. Sci.* 2000 **9**, 1045.

STRUCTURAL EVIDENCE FOR SUBSTRATE ASSISTED CATALYTIC MECHANISM OF BEE VENOM HYALURONIDASE, A MAJOR ALLERGEN OF BEE VENOM

Z. Marković-Housley and T. Schirmer

Division of Structural Biology
Biozentrum, University of Basel
4056 Basel, Switzerland

1 INTRODUCTION

Bee venom hyaluronidase (Hya) is a hydrolase which specifically cleaves the β-1,4 glycosidic bond of hyaluronic acid (HA) in the cell surface of skin. HA is the most abundant glycosaminoglycan of the vertebrate extracellular spaces and is found in virtually all tissues and body fluids. It is a high molecular weight linear polymer built from repeating disaccharide units D-glucuronic acid(β-1,3)–N-acetyl-D-glucosamine(β-1,4) (GlcA-GlcNAc), Figure 1.

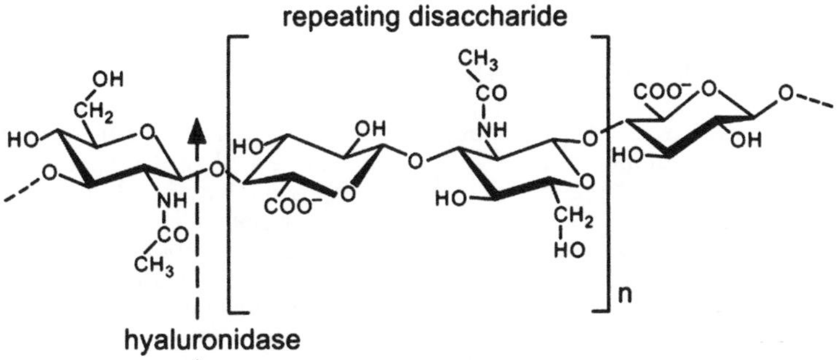

Figure 1: *Hyaluronic acid*

At physiological conditions HA is a large, charged and extended polysaccharide which is exceptionally hydrophilic. The ability to bind large amounts of water confers to HA solution special viscoelastic properties which are the basis of its structural role as a stabilizer, joint lubricant and shock absorbent[1,2]. Highly hydrated HA also promotes cell locomotion and migration which are very prominent in embryogenesis, wound healing and malignancy, processes characterized by rapid increase of HA.

Hyaluronidases, the enzymes which degrade hyaluronic acid, are widely spread in nature and have a wide range of pH optima. They have been classified in three major groups based on their mechanism and substrate specificity, deduced from biochemical analysis of enzymes and their reaction products[3]. Hya belongs to group 1 represented by mammalian-type hyaluronidases (EC 3.2.1.35) found in testis and lysosomes, as well as in the venom of bees, wasps, snakes and scorpions. The enzymes of this group are endo-β-N-acetyl-D-hexosaminidases which specifically cleave the internal β-1,4 glycosidic linkage between GlcNAc and GlcA of HA and 4-S and 6-S chondroitin sulfate producing a non reactive tetrasaccharide as the main end product. Besides hydrolysis, it is known that a testicular enzyme also catalyzes the reverse reaction, transglycosylation[4,5]. Consequently, processing of HA hexamers leads to a mixture of oligosaccharides, comprising not only tetrasaccharides, but also a substantial proportion of octa-, hexa- and small amount of di-saccharides. The turnover of HA in vertebrates is regulated by the combined action of Hya and two exoglycosidases, β-glucuronidase and β-N-acetyl glucosaminidase. Group 2 hyaluronidase is represented by leech enzyme (EC 3.2.1.36) which acts as endo-β-D-glucuronidase whereas Group 3 comprises bacterial HA-lyases (EC 4.2.99.1) that act as endo-N-acetyl D-hexosaminidase by β elimination.

Hya and mammalian hyaluronidases belong to glycosyl hydrolase family 56[6,7] and the recently reported structure of Hya[8] is the first representative of this family. Bee venom hyaluronidase shares 30% sequence similarity with the sperm PH-20 protein[9], involved in fertilization, and the human lysosomal enzymes Hyal-1[10] and Hyal-2[11] which regulate HA turnover. The PH-20 protein, GPI-bound to the acrosomal membrane of the sperm, degrades the HA-rich matrix in which cumulus cells surrounding the egg are embedded, enabling sperm-egg adhesion[12].

Hyaluronidase and phospholipase A2 are two major allergens of bee venom which can induce serious, occasionally fatal, allergic reactions in humans. The knowledge of the Hya structure is the first step towards a better understanding of the mechanism underlying allergic reaction and, hopefully, towards a development of an improved treatment of bee venom induced allergy. Bee venom and human hyaluronidases show conservation of the active site residues[8], suggesting that the mammalian enzymes may operate *via* a catalytic mechanism similar to that of Hya, presented here. The structure of Hya will be used to model the sequence related mammalian enzymes whose malfunctioning is related to many diseases.

2 BIOCHEMICAL CHARACTERIZATION OF HYA

Native hyaluronidase, isolated from honey bee venom, is a single chain secreted protein composed of 350 amino acids with M_r of 43000 and pI of 8.7[13]. The secreted protein is derived from a precursor which is composed of a signal peptide and a short pro-segment. The cDNA nucleotide sequence of cloned Hya is known[9]. The mature enzyme contains four potential glycosylation sites (carbohydrate content of 7%) and two disulfide bridges. Recombinant His-tagged Hya has been expressed in prokaryotic (*E. coli*) and eukaryotic (*Baculovirus*) hosts and purified by Ni-chelation chromatography[14]. Only the *Baculovirus*-expressed enzyme has enzymatic activity and IgE-binding capacity similar to native Hya, and was used for crystallization and 3D structure determination.

3 THE OVERALL FOLD

The structure of unliganded Hya has been solved in two different crystal forms, monoclinic and trigonal at resolutions of 1.6 and 2.1 Å, respectively[8]. These structures revealed an unusual overall fold, comprised of a 7 stranded β barrel (β1–β7) surrounded by 10 α helices (A–J), Figure 2. The most striking feature of the $(\beta/\alpha)_7$ barrel is the large gap (7–8 Å) between strands 1 and 2 and a prominent groove formed by the loops at the C-terminal end of β barrel, suited for substrate binding. The groove is large enough (30×10 Å) to accommodate a HA-hexamer or octamer and is lined by many conserved residues, e.g. the active site residues Asp111 and Glu113, Arg116 and Arg244 and many aromatic residues as usually found in other

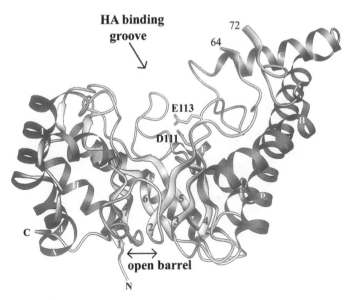

Figure 2: *Ribbon representation of the three dimensional structure of Hya. Side view of the (β/α)₇ barrel composed of β strands 1–7 (light gray) and α helices A–J (dark gray). The active site residue Glu113 and Asp111 are shown as stick models.*

sugar binding proteins.

The 7-stranded "open" barrel has also been observed in the structure of quinolinic acid phosphoribosyltransferase (QAPRTase)[15], in which a gap is found between strands 2 and 3 and is partially occupied by the substrate. The fold of cellobiohydrolase II[16] and the related thermophilic endocellulase E2[17] is also a seven stranded (β/α) barrel, although they contain equally spaced strands. A search for similar folds showed that the top scoring proteins (Z = 11–12) all have a regular (β/α)₈ barrel fold, e.g. β-amylases from *Bacillus cereus*[18] and *Soybean*[19], β-glucuronidase[20], myrosinase[21], endo-β-xylanase[22]. Similarity of the active site architecture was only observed with the bacterial chitinase A[23] which had a score of Z = 9 but showed a similar arrangement of the two catalytic acids and several conserved aromatic residues, Figure 3.

4 THE STRUCTURE OF HYA/HA-TETRAMER COMPLEX

The complex of Hya and HA-tetramer was obtained by co-crystallization of enzyme with modified HA hexamer, which had fluorescein attached to the C1 atom of GlcNAc at the reducing end. Solution studies have shown that the testicular enzyme cannot process this compound[5]. The structure of the Hya complex in the trigonal crystal form was solved at 2.65 Å resolution and showed, unexpectedly, a tetrasaccharide (GlcA-GlcNAc)₂ bound in the subsites −4 to −1 of the substrate binding groove, with the reducing end GlcNAc in subsite −1 positioned close to Glu113, the putative proton donor. Evidently, the tetrasaccharide is produced by hydrolysis of β-1,4 bond joining saccharides GlcNAc −1 and GlcA +1, while the leaving group (disaccharide

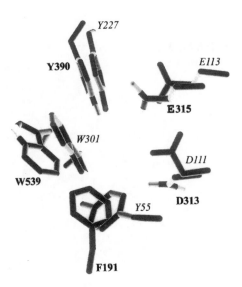

Figure 3: *Superposition of the active site residues of Hya (labeled in italic) and chitinase A from Serratia marcescens (labeled in boldface). Shown are the catalytic residues D111 and E113 of Hya and D313 and E315 of chitinase A, as well as several conserved aromatic residues.*

with attached fluorescein) is removed. The tetrasaccharide is most tightly bound at the reducing end where it forms hydrogen bonds with Glu113, Asp111 and Tyr227 and hydrophobic interactions with Trp301 and Tyr184, Figure 4. In contrast, GlcA in subsite −4 is loosely bound. The GlcNAc residue in subsite −3 is hydrogen bonded to the hydroxyl group of a non conserved Ser304 whereas in subsite −2 the carboxyl group of a GlcA residue forms two hydrogen bonds with the hydroxyl group and main chain amide of a non conserved Ser303, and its sugar ring stacks against Tyr55.

5 THE CATALYTIC MECHANISM

In general, glycosidases act via a double or single nucleophilic displacement mechanism which results in either the retention or inversion of the anomeric carbon configuration, respectively. In both cases the scissile glycosidic bond is positioned between the carboxylates of the two catalytic acids, one acting as acid/base and the other as the nucleophile. In Hya, the only two acidic residues found in the binding groove are Asp111 and Glu113, at the C-terminal end of strand β3. They are strictly conserved in insect and mammalian enzymes and are joined by a short hydrogen bond (2.5 Å), in both unliganded and complexed form, Figure 4. The catalytic acid in Hya is Glu113 since it is bound to the glycosidic oxygen of GlnNAc in subsite −1 whereas one of the many roles of the proximate Asp111 is to keep the side chain of Glu113 in the proper orientation for catalysis. This is in agreement with zero enzymatic activity found upon substituting Glu113 by Gln in Hya (unpublished results) and human sperm PH-20 protein[24] and 3% activity for the Asp111Asn mutant of PH-20[24].

In Hya, the proteinaceous nucleophile is missing since, for steric reasons, Asp111 is not in the position to act as a nucleophile and there is no other suitable protein nucleophilic group in the vicinity. Instead, the carbonyl oxygen of the acetyl side

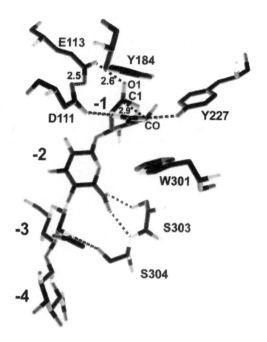

Figure 4: *Active site of Hya showing interactions of protein residues directly involved in binding of HA tetramer GlcA-GlcNAc-GlcA-GlcNAc.*

chain is positioned ideally to act as a nucleophile, at hydrogen bonding distance from the anomeric C1 atom, Figure 4. Several interactions enable the catalytically favourable orientation of the N-acetyl side chain which is a prerequisite for the catalysis to occur: (i) two hydrogen bonds, one between its amide NH and Asp111 and the other between its carbonyl oxygen and the hydroxyl group of Tyr227, and (ii) hydrophobic interactions between the methyl group of the N-acetyl side chain with Tyr184 and Trp301. Consequently, the acetyl side chain is rotated below the −1 sugar ring plane so that its carbonyl oxygen O7 is only 2.9 Å away from the anomeric C1 atom, which is found in the β conformation typical for a retaining enzyme. Moreover, the saccharide ring at −1 is deformed from the usual $^4C_1$ chair observed in subsites −2 to −4, to a $^{4,1}C$ boat conformation. While the resolution of our complex structure (2.65 Å) does not distinguish between $^{4,1}C$ boat and other closely related conformations, or mixture of these, it is clear that GlcNAc −1 does not adopt the $^4C_1$ chair conformation. With an undistorted sugar ring in subsite −1, it would be impossible, by a simple rotation around C2-N2, to bring the acetamido group in a position similar to that shown in Figure 4, which would allow favorable hydrogen bonding interactions with Tyr227 and Asp111. Distortion of sugar −1 facilitates the cleavage reaction by bringing the glycosidic oxygen nearer to Glu113 and the carbonyl oxygen O7 close to the anomeric C1 carbon.

The interactions between Hya and GlcNAc in subsite −1 are similar to those observed in the structure of chitobiase (family 20) in the abortive complex with its substrate disaccharide chitobiose[25]. Chitobiase also lacks a protein nucleophile and acts via the substrate's acetamido group participation. Upon binding of chitobiose, a O7-C1 distance of 2.8 Å is achieved as well as the distortion of the tightly bound saccharide −1 into a twisted boat conformation. There is remarkable similarity in the pattern of hydrogen bonding and hydrophobic interactions involved in docking the sugar −1 in our enzyme/product complex and that in the enzyme/substrate complex of chitobiase, Table 1. Similarity is also observed in the complexes of plant hevamine (family 18) with chitotetraose[26], Table 1, and with a natural inhibitor allosamidine[27]. In contrast to Hya, saccharide −1 in the hevamine/chitotetraose complex is found in the usual $^4C_1$ chair conformation.

Table 1: *Conserved interactions of saccharide −1 in the sugar complexes with Hya (family 56), chitobiase (family 20) and hevamine (family 18), which operate via a substrate assisted mechanism.* * *($\leftrightarrow$ and $\cdots$ stand for hydrophobic and hydrogen bonding interactions, respectively.)*

Protein/ Sugar Complex	Hyaluronidase/ HA-tetramer	Chitobiase/ Chitobiose	Hevamine/ Chitotetraose
PDB code	1FCV	1QBB (10)	nd [†]
Protein/sugar −1 (Å)			
E-$O^\epsilon\cdots$O1 [‡]	E113$\cdots$O1 (2.6)	E540$\cdots$O1 (2.8)	E127$\cdots$O1 (3.0)
D-$O^\delta\cdots$NH [‡]	D111$\cdots$NH (2.8)	D539$\cdots$NH (2.9)	D125$\cdots$NH (3.0)
Y-OH $\cdots$O7	Y227$\cdots$O7 (2.8)	Y669$\cdots$O7 (3.3)	Y183$\cdots$O7 (2.75)
Y(W) $\leftrightarrow$ CH$_3$ [§]	Y184 $\leftrightarrow$ CH$_3$	W616 $\leftrightarrow$ CH$_3$	Y6 $\leftrightarrow$ CH$_3$
W $\leftrightarrow$ CH$_3$ [§]	W301 $\leftrightarrow$ CH$_3$	W737 $\leftrightarrow$ CH$_3$	W255 $\leftrightarrow$ CH$_3$
Sugar −1			
C1$\cdots$O7 (Å) [¶]	2.9	2.8	2.8
conformation	$^{1,4}C$ boat	4 sofa	$^4C_1$ chair

* Glu is a proton donor and acetamido O7 oxygen of sugar −1 is a nucleophile.

[†] Coordinates not deposited. See paper by Tews et al.[26]

[‡] Asp precedes proton donor Glu; both acids are important for catalysis.

[§] Hydrophobic interactions between protein aromates and acetamido CH$_3$ group.

[¶] Protein/sugar −1 interactions bring the O7 oxygen close to anomeric C1 atom.

The role of the proximate Asp111 is essential in providing the catalytically competent orientation of the Glu113 and the N-acetyl side chains, which both enable hydrolysis and the retention of C1 atom configuration of the product. A similar role is performed by the equivalent Asp residue in the enzymes of family 18. In chitinolytic enzyme and Hya, the Asp-Glu pair is found at the C-terminal end of barrel strands $\beta4$ and $\beta3$, respectively, Table 2. However, these two strands are topologically equivalent given the fact that the 7-stranded β barrel of Hya is missing strand 2, and thus, $\beta3$ of Hya corresponds to $\beta4$ in chitinolytic enzymes, which are $(\beta/\alpha)_8$ barrels.

Our results, derived from the structure of a Hya/HA-product complex, support the conclusion that Hya (family 56) is a retaining glycosidase which operates via a substrate assisted mechanism similar to that proposed for chytinolytic enzymes of the glycosyl hydrolase family 18 and 20[27,28]. This double displacement mechanism involves the binding of the saccharide −1 in a boat conformation and a formation and subsequent hydrolysis of a covalent oxazolinium ion intermediate[28], Figure 5. In Hya, the proton donor is Glu113 and the nucleophile is the acetamido oxygen of the substrate Figure 5. Upon donation of the proton to the glycosidic oxygen the β-1,4 bond between GlcNAc −1 and GlcA +1 is cleaved and the +1 sugar group leaves the active site. The close proximity of the negatively charged acetamido oxygen O7 and the positively charged anomeric C1 lead to the stabilization of the transition state by the formation of a covalent oxazolinium intermediate, as postulated for retaining β glycosyl hydrolases of family 18 on the basis of structural data[26,28,29] as well as

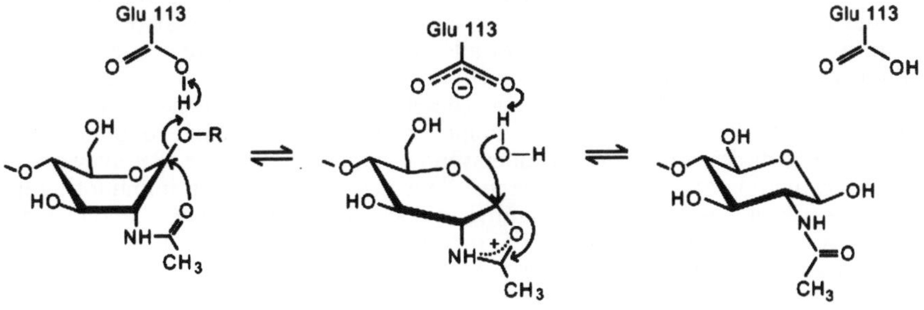

Figure 5: *Double-displacement substrate assisted mechanism of bee venom hyaluronidase. Saccharide −1 binds in the boat conformation (1) and catalysis is proposed to occur via a formation of covalent cyclic oxazolinium ion intermediate (2).*

theoretical calculations[30]. Subsequent hydrolysis of the oxazolinium ion intermediate, via a transition state having oxocarbenium ion character (not shown), releases the product in the usual chair conformation[28] Figure 5. Adversely, in Hya/HA-product complex the sugar −1 is found in a boat conformation typical for an enzyme/substrate complex which is due to strong interactions between saccharide −1 and the protein.

6 PERSPECTIVES

Single mutants of the strictly conserved active site carboxylates (Glu113 and Asp111) were constructed (Glu113Gln and Asp111Asn) with the goal of trapping the enzyme-substrate complex, since the mutated proteins should be able to bind the HA hexamer or octamer without processing. The structures of these mutants will help to elucidate details of catalytic mechanism and structural adjustments of the protein upon substrate binding.

The Hya structure will be used to model the homologous human sperm PH-20 protein and lysosomal hyaluronidases, Hyal1 and Hyal2. The knowledge of these structures and their catalytic mechanisms could make a significant impact in the treatment of disorders related to defective functioning of these enzymes, such as in

Table 2: *Catalytic Asp-Glu pair in Hya and chitinolytic enzymes which use Glu as protein donor but lack the proteinaceous nucleophile.*

Enzyme	Family	C-end of $\beta4$	H-bonded (Å)
Endo H	18	...F **D D E**	
Endo F1	18	...F **D D E**	D130···E132 (2.5)
Hevamine	18	...F **D I E**	D125···E127 (2.6)
Chitinase A	18	...I **D W E**	
Hyaluronidase	56	...I **D F E**	D111···E113 (2.5)

osteo- and rheumatoid arthritis, probably by the structure-based design of specific inhibitor(s).

Hya is a potent major allergen of bee venom and the knowledge of the structural determinants responsible for the allergenic potency of Hya is expected to have important clinical implications. A promising approach is to inhibit the onset of an allergic reaction, which is triggered by binding of allergen to specific patient IgE-antibodies. The full characterization of the antigen-antibody recognition sites will require the elucidation of the structure of complexes between Hya and Fab fragments (from various monoclonal antibodies against Hya), combined with biochemical and immunochemical studies. Site directed mutagenesis of epitope residues may lead to mutant variants with low IgE binding activity (hypoallergens[31]) which can be used in immunotherapy without the risk of side effects and anaphylactic shock. Alternatively, the knowledge of the epitope structure may lead to the design of monovalent ligands which would block the cross-linking of IgE receptors[32].

References

1. T. C. Laurent, *The biology of hyaluronan*, Ciba Foundation Symp 143, John Wiley & Sons, New York, 1989.
2. T. C. Laurent and J. R. E. Fraser, *FASEB J.*, 1992, **6**, 2397.
3. K. Meyer, Hyaluronidases, in *The Enzymes, 3rd ed.*, edited by P. D. Boyer, volume V, pages 307–320, Academic Press, New York, 1971.
4. J. A. Cramer, L. C. Bailey, C. A. Bailey, and R. T. Miller, *Biochim. Biophys. Acta*, 1994, **1200**, 315.
5. K. Takagaki, T. Nakamura, J. Izumi, H. Saitoh, M. Endo, K. Kojima, I. Kato, and M. Majima, *Biochemistry*, 1994, **33**, 6503.
6. B. Henrissat, *Biochem. J.*, 1991, **280**, 309.
7. B. Henrissat and A. Bairoch, *Biochem. J.*, 1996, **316**, 695.
8. Z. Marković-Housley, G. Miglierini, L. Soldatova, P. J. Rizkallah, U. Müller, and T. Schirmer, *Structure*, 2000, **8**, 1025.
9. M. Gmachl and G. Kreil, *Proc. Natl. Acad. Sci. USA*, 1993, **90**, 3569.
10. G. I. Frost, T. B. Csoka, T. Wong, and R. Stern, *Biochem. Biophys. Res. Commu.*, 1997, **236**, 10.
11. G. Lepperdinger, B. Strobl, and G. Kreil, *J. Biol. Chem.*, 1998, **273**, 22466.
12. P. Primakoff, W. Lathrop, L. Woolman, A. Cowan, and D. Myles, *Nature*, 1988, **335**, 543.
13. D. M. Kemeny, N. Dalton, A. J. Lawrence, F. L. Pearce, and C. A. Vernon, *Eur. J. Biochem.*, 1984, **139**, 217.
14. L. N. Soldatova, R. Crameri, M. Gmachl, D. Kemeny, M. Schmidt, M. Weber, and U. Mueller, *J. Allergy Clin. Immunol.*, 1998, **101**, 691.
15. J. C. Eads, D. Ozturk, T. B. Wexler, C. Grubmeyer, and J. C. Sacchettini, *Structure*, 1997, **5**, 47.
16. J. Rouvinen, T. Bergfors, T. Teeri, J. K. Knowles, and T. A. Jones, *Science*, 1990, **249**, 380.
17. M. Spezio, D. B. Wilson, and P. A. Karplus, *Biochemistry*, 1993, **32**, 9906.
18. B. Mikami, M. Adachi, T. Kage, E. Sarikaya, T. Nanmori, R. Shinke, and S. Utsumi, *Biochemistry*, 1999, **38**, 7050.
19. B. Mikami, M. Degano, E. J. Hehre, and J. C. Sacchettini, *Biochemistry*, 1994, **33**, 7779.
20. S. Jain, W. B. Drendel, Z. Chen, F. S. Mathews, W. S. Sly, and J. H. Grubb, *Nature Struct. Biol.*, 1996, **3**, 375.

21. W. P. Burmeister, S. Cottaz, H. Driguez, R. Iori, S. Palmieri, and B. Henrissat, *Structure*, 1997, **5**, 663.
22. L. Lo Leggio, S. Kalogiannis, M. K. Bhat, and R. W. Pickersgill, *Proteins*, 1999, **36**, 295.
23. A. Perrakis, I. Tews, Z. Dauter, A. B. Oppenheim, Chet, I., K. S. Wilson, and C. E. Vorgias, *Structure*, 1994, **2**, 1169.
24. S. Arming, B. Strobl, W. C., and G. Kreil, *Eur. J. Biochem.*, 1997, **247**, 810.
25. I. Tews, A. Perrakis, A. Oppenheim, Z. Dauter, K. S. Wilson, and C. E. Vorgias, *Nature Struct. Biol*, 1996, **3**, 638.
26. I. Tews, A. C. Terwisscha van Scheltinga, A. Perrakis, K. S. Wilson, and B. W. Dijkstra, *J. Am. Chem. Soc.*, 1997, **119**, 7954.
27. A. C. Terwisscha van Scheltinga, S. Armand, K. Kalk, A. Isogai, B. Henrissat, and B. W. Dijkstra, *Biochemistry*, 1995, **34**, 15619.
28. S. Drouillard, S. Armand, G. Davies, C. Vorgias, and H. B., *Biochem. J.*, 1997, **328**, 945.
29. A. White and D. Rose, *Curr Opin Struct Biol*, 1997, **7**, 645.
30. K. Brameld, W. Shrader, B. Imperiali, and W. r. Goddard, *J. Mol. Biol.*, 1998, **2809**, 913.
31. F. Ferreira, K. Hirtenlehner, A. Jilek, J. Godnik-Cvar, H. Breiteneder, R. Grimm, K. Hoffmann-Sommergruber, O. Scheiner, D. Kraft, M. Breitenbach, H.-J. Rheinberger, and C. Ebner, *J. Exp. Med.*, 1996, **183**, 599.
32. R. Valenta, S. Vrtala, T. Ball, S. Laffer, P. Steinberger, and D. Kraft, *ACI News*, 1994, **6/6**, 165.

STRUCTURE AND FUNCTION OF CLASS I α1,2-MANNOSIDASES INVOLVED IN GLYCOPROTEIN BIOSYNTHESIS

A. Herscovics[1], F. Lipari[1], B. Sleno[1], P.A. Romero[1], F. Vallée[2], P. Yip[2] and P. L. Howell[2]

[1]McGill Cancer Centre, McGill University, Montréal, Canada
[2]The Hospital for Sick Children, Toronto, Canada.

1 INTRODUCTION

Class I α1,2-mannosidases (glycosylhydrolase family 47) are a family of proteins that play an important role in the early stages of glycoprotein synthesis (1-3).

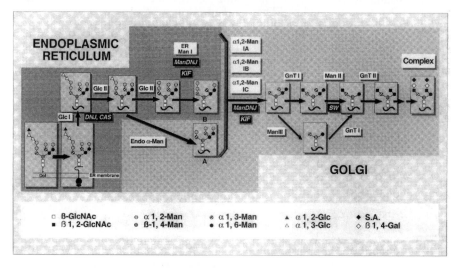

Figure 1 *N-glycan processing in mammalian cells – ER ManI, α1,2-Man IA, IB and IC are Class I α1,2-mannosidases* (modified from (3))

During N-glycan formation, a $Glc_3Man_9GlcNAc_2$ oligosaccharide precursor is synthesized attached to dolichol pyrophosphate in the endoplasmic reticulum (ER) (Figure 1). This preformed oligosaccharide is then transferred to asparagine residues onto nascent polypeptide chains and is subsequently modified by ER and Golgi glycosidases and by Golgi glycosyltransferases. The glucose residues are removed by glucosidases I and II in the ER, and some mannose is removed by ER α1,2-mannosidase I that primarily forms $Man_8GlcNAc_2$

isomer B. There is increasing evidence for a role of this ER α1,2-mannosidase and for an enzymatically inactive α1,2-mannosidase homolog in ER quality control leading to the degradation of misfolded glycoproteins by the proteasome.

In mammalian cells, additional α1,2-linked mannose residues are removed by Golgi α1,2-mannosidases IA, IB or IC, derived from different genes and expressed in a cell-specific manner. Trimming by the Class I α1,2-mannosidases is essential for further maturation of N-glycans to complex and hybrid structures since the resulting Man$_5$GlcNAc$_2$ is a substrate for the first Golgi glycosyltransferase. Subsequently, additional glycosyltransferases and α-mannosidases belonging to another protein family participate in the elaboration of the variety of oligosaccharide structures present on mammalian glycoproteins.

In *Saccharomyces cerevisiae*, the ER α1,2-mannosidase that forms Man$_8$GlcNAc$_2$ isomer B is the only α1,2-mannosidase involved in N-glycan biosynthesis. Mannose residues are then added by different mannosyltransferase in the yeast Golgi.

2 CLASS I α1,2-MANNOSIDASES

Class I mannosidases are type II membrane proteins of 63-80kDa that require Ca^{+2} for activity. They specifically cleave α1,2-linked mannose residues but they do not act on aryl α-mannopyranosides. They are inhibited by pyranose analogs such as 1-deoxymannojirimycin (ManDNJ) and kifunensine (KIF). There are presently more than 50 known sequences from different species belonging to this protein family. Only a few of these have been expressed as recombinant proteins for detailed structure-function studies.

2.1 Yeast ER α1,2-Mannosidase

This enzyme has served as a model of Class I α1,2-mannosidases. It is a 63kDa membrane protein consisting of an N-terminal transmembrane domain followed by a large C-terminal catalytic domain that faces the lumen of the ER. Its catalytic domain was expressed in mg quantities as a soluble protein secreted into the medium of *Pichia pastoris* by replacing the transmembrane domain with a cleavable signal sequence. It was demonstrated by high resolution proton NMR to be an inverting glycosidase. Analysis of tryptic peptides and mutagenesis studies showed that it has two disulfide bonds and a free sulfhydryl group. One of the disulfide bonds is essential for enzyme activity whereas the free sulfhydryl is not required for activity. The catalytic domain has nine conserved acidic amino acids that were shown by mutagenesis to all be required for enzymatic activity (4,5).

2.2 Structure of the Yeast ER α1,2-Mannosidase

It is the first Class I α1,2-mannosidase whose structure has been determined by X-ray crystallography (1.54Å resolution) (6). The yeast enzyme is a glycoprotein with three N-glycans. Many different constructs were prepared to obtain reproducible crystals that diffracted at high resolution. These attempts involved several strategies. Carbohydrate groups were removed enzymatically or by mutagenesis. Truncations of the N-terminus were performed to eliminate N-terminal heterogeneity due to proteolysis. Finally, five amino acids were deleted from a putative loop to eliminate an internal site of proteolytic cleavage. The best crystals were obtained with the fully glycosylated protein containing the reconstructed loop and beginning at amino acid #34.

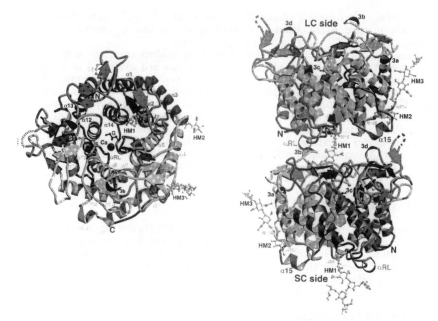

Figure 2 *Yeast α1,2-mannosidase structure.* Left, Axial view of a single molecule from the LC side. Right, Two adjacent molecules viewed at 90° to the orientation on the left. (reproduced from (6) with permission from Oxford University Press)

The yeast ER α1,2-mannosidase structure consists of an $(\alpha\alpha)_7$ barrel with consecutive helices alternating from the inside to the outside of the barrel. The structure is not an open channel as one side of the barrel is plugged by a β-hairpin arising from the C-terminus of the protein and creating a cavity of ~15 Å in depth. The Ca^{+2} ion as well as the conserved acidic amino acids and a molecule of glycerol (G) used as cryoprotectant are found above this hairpin at the bottom of the barrel where the catalytic site is located (Figure 2).

The yeast ER α1,2-mannosidase catalytic domain structure includes the three predicted N-glycans extending from the surface of the protein. Only 2-3 sugar residues are observed for two of these N-glycans (HM2, HM3) but the third N-glycan HM1 is stabilized by protein-carbohydrate interactions between adjacent molecules within the crystal. At this site a $Man_5GlcNAc_2$ structure is seen to extend from one molecule into the barrel of the adjacent symmetry-related molecule (Figure 2). The structure represents an enzyme-product complex since it is missing the mannose residue that is cleaved by the enzyme. The protein-carbohydrate interactions and shortening of the loop (RL) located at the interface between adjacent protein molecules facilitated optimum crystal formation.

2.3 Effect of Mutagenesis on the Specificity of the Yeast ER α1,2Mannosidase

The protein-carbohydrate interactions visualized in the structure show that among many amino acids close to the carbohydrate moiety, Arg^{273} forms the largest number of contacts with several mannose residues (Figure 3). This particular arginine is conserved in the human ER α1,2-mannosidase that has the same specificity in forming $Man_8GlcNAc_2$ isomer B from $Man_9GlcNAc_2$. It is replaced by leucine in the mammalian Golgi α1,2-mannosidases of different specificity that trim $Man_9GlcNAc_2$ to $Man_8GlcNAc_2$ isomers A and/or C intermediates toward the formation of $Man_5GlcNAc_2$. To gain some insight into the structural determinants of these different specificities, Arg^{273} was mutated to leucine (7). The R273L mutant has a unique specificity and readily converts $Man_9GlcNAc_2$ to $Man_5GlcNAc_2$(Figure 3), but the order of removal of the $\alpha1,2$-linked mannose residues is different from that observed with the Golgi enzymes since $Man_8GlcNAc_2$ isomer B is an intermediate.

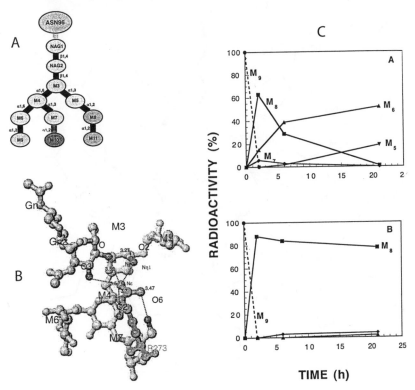

Figure 3 *Role of R273 in specificity of the yeast α1,2-mannosidase.* A. $Man_9GlcNAc_2$. The following mannose residues are not visible in the structure: M10 (mannose cleaved by the enzyme), M8, M9, and M11.
B. Interactions of R273 with the oligosaccharide showing distance in Å.
C. Trimming of $Man_9GlcNAc$ (M9) by the R273L mutant (A) and by the wild-type enzyme (B). (Reproduced from (7) with permission from The American Society for Biochemistry and Molecular Biology).

2.4 Structure of the Yeast ER α1,2-Mannosidase Complex wih Inhibitor

The structure of the enzyme complex with the mannose analog 1-deoxymannojirimycin (ManDNJ) was also determined by X-ray crystallography (1.59Å). The inhibitor binds to the active site at the position of the mannose specifically cleaved by the enzyme. There is no major conformational change observed upon inhibitor binding. The Ca^{+2} is 8-fold coordinated, interacting with the O2' and O3' hydroxyls of the ManDNJ ring and with the conserved acidic residues indirectly *via* water molecules (Figure 4).

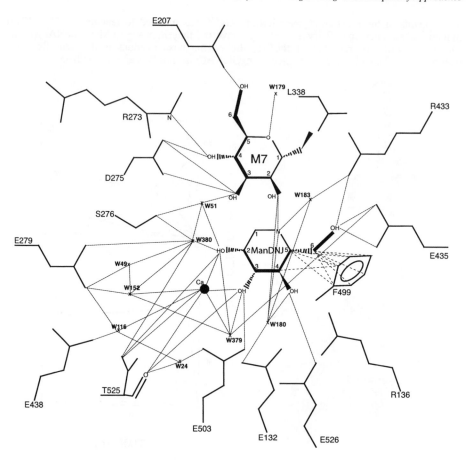

Figure 4 *Binding of 1-deoxymannojirimycin(dMNJ) to the yeast α1,2-mannosidase.*
H-bonds are shown by thin lines and van der Waals interactions by dashed lines. All
conserved acidic amino acids required for activity are shown.

The ManDNJ ring is stabilized in a $^1C_4$ conformation, with its C-1 position in place of the
mannose cleaved by the enzyme and its C-4, C-5 and C-6 overlapping with the glycerol
molecule. There are three acidic residues (E132 on one side of the cleaved bond, E435 and
D275 on the other side) that are located at an appropriate distance to participate in catalysis
(Figure 5). The structure of the yeast α1,2-mannosidase and of the human ER α1,2-
mannosidase that was determined by molecular replacement suggest a novel catalytic
mechanism for these inverting glycosidases (8).

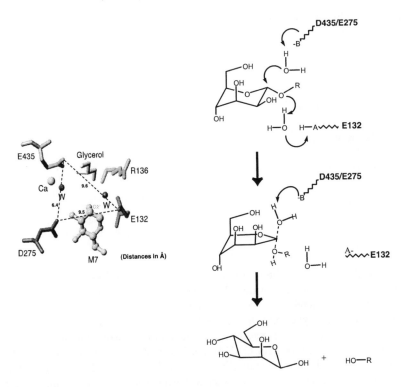

Figure 5 *Proposed catalytic mechanism.* (modified from (8) with permission of The American Society for Biochemistry and Molecular Biology).

References

1. R. Kornfeld and S. Kornfeld, *Annu. Rev. Biochem.*, 1985, **54**, 631-664.
2. A. Herscovics, *Biochim. Biophys. Acta*, 1999, **1426**, 275-285.
3. A. Herscovics, *Biochim. Biophys. Acta*, 1999, **1473**, 96-107.
4. F. Lipari and A. Herscovics, *J. Biol. Chem.*, 1996, **271**, 27615-27622.
5. F. Lipari and A. Herscovics, *Biochemistry,* 1999, **38**, 1111-1118.
6. F.Vallée, F. Lipari, P.Yip, B. Sleno, A. Herscovics and P. L. Howell, *EMBO J.*, 2000, **19**, 581-588.
7. P. A. Romero, F.Vallée, P. L. Howell and A. Herscovics, *J. Biol. Chem.*, 2000, **275**, 11071-11074.
8. F.Vallée, K. Karaveg, A. Herscovics, K. W. Moremen and P. L. Howell, *J. Biol. Chem.*, 2000, **275,** 41287-41298.

Acknowledgments

This work was supported by grants from NIH (GM31265), NSERC and CIHR

STRUCTURE AND FUNCTION OF LYTIC TRANSGLYCOSYLASES FROM
Pseudomonas aeruginosa

N. T. Blackburn and A. J. Clarke

Department of Microbiology
University of Guelph
Guelph, Ontario N1G 2W1 Canada

1 INTRODUCTION

As with plants, algae, and fungi, bacteria produce a rigid polymer of $\beta,1\rightarrow4$ linked glycosides that comprises the bulk of their cell walls. These polymers withstand the internal turgor pressure of the cytoplasm thereby maintaining the integrity of the cytoplasmic membranes. However, unlike cellulose, xylan, and chitin, which are homopolymers of glucose, xylose, and *N*-acetylglucosamine (GlcNAc), respectively, the peptidoglycan (or murein) of bacteria is a heteropolymer of two amino sugars, GlcNAc and *N*-acetylmuramic acid (MurNAc), the latter of which is modified with a peptide (Figure 1). This stem peptide is comprised of alternating L- and D- amino acids which may vary with genera and species. Chains of these alternating aminosugars are thought to form "hoops" around the circumference of the cell, which in the Gram-negative bacteria, may only be one layer thick. For this reason, the carbohydrate polymers need to be crosslinked together in order to provide the required rigidity of the wall polymer. This crosslinking is achieved through neighbouring stem peptides where, typically, the fourth amino acid residue (usually D-alanine) of one is covalently attached to the third residue (usually a diamino acid such as L-lysine or diaminopimelic acid) of the other. Thus, the peptidoglycan of a bacterium is one large macromolecule, or sacculus, which maintains the integrity and shape of the cell.

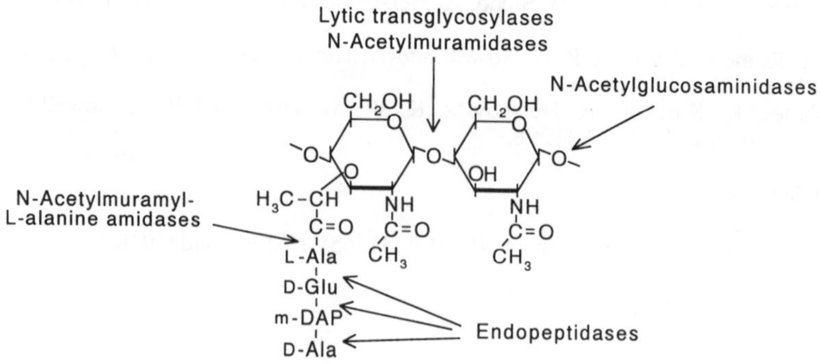

Figure 1 *Structure of peptidoglycan repeating unit and sites of its enzymatic cleavage*

Given the importance of peptidoglycan to a bacterium, a number of its biosynthetic enzymes have been exploited as the target for antibacterial agents. For example, the β-lactam antibiotics (penicillins, cephalosporins, monobactams) which block the crosslinking of peptidoglycan have proven to be an extremely valuable therapeutic against bacterial pathogens over the past 60 years. However, many important pathogens have developed resistance to the commonly used drugs, and so recently much effort has been made to find alternative targets for the development of new antibiotics. In this chapter, we review the current understanding of one of these potential targets, the lytic transglycosylases.

2 FUNCTION AND ACTIVITY OF LYTIC TRANSGLYCOSYLASES

The biosynthesis of peptidoglycan occurs in all three compartments of a bacterial cell: the cytoplasm, on the cytoplasmic membrane, and in the periplasm. The soluble precursor molecule, UDP-linked MurNAc-pentapeptide is synthesized from UDP-GlcNAc, phosphoenolpyruvate, and amino acids and it is then transferred to the membrane carrier molecule bactoprenol (undecaprenyl phosphate) to form Lipid I. A glycosidic linkage to GlcNAc yields Lipid II, and the β-1,4-linked GlcNAc-MurNAc-pentapeptide is then translocated across the cytoplasmic membrane to be added into the growing peptidoglycan sacculus. These latter stages of peptidoglycan biosynthesis thus occur outside the cell and involve the penicillin-binding proteins, which catalyze both transglycosylation and transpeptidation reactions.

To provide new sites for the incorporation of synthesis, the peptidoglycan has to be strategically cleaved without loss of cell integrity. Given the complex structure of peptidoglycan involving both glycosidic and peptide bonds, a variety of different enzymes are produced to achieve this (Figure 1). One such class of enzymes are the lytic transglycosylases (LTs) which act at the same site as muramidases (lysozymes), that is the β-1,4 linkage between MurNAc and GlcNAc residues. Unlike the muramidases, however, the LTs are not hydrolases but instead cleave the linkage with the concomitant formation of 1,6-anhydromuramoyl residues (anhMurNAc) (Figure 2).

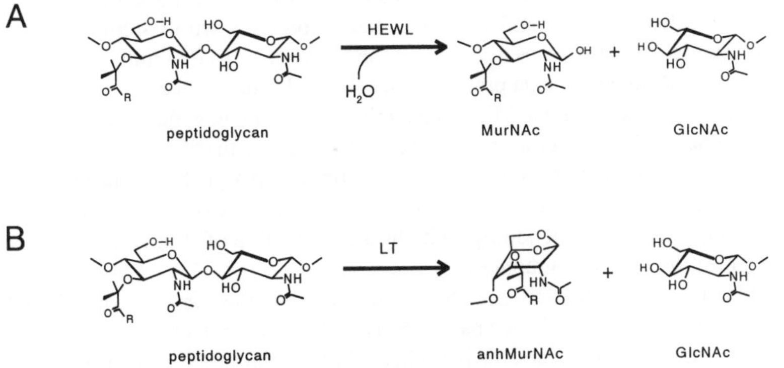

Figure 2   *Cleavage reactions performed on peptidoglycan by (A) muramidases such as hen egg-white lysozyme (HEWL) and (B) lytic transglycosylases (LT)*

Escherichia coli produces at least six different LTs, some of which are membrane-bound (Mlt) lipoproteins while others are released in soluble (Slt) form (reviewed in ref. 1). Five of these six function as exo-acting enzymes, releasing anhydro-disaccharide-peptide units from either the reducing or non-reducing ends of peptidoglycan chains. The sixth, EmtA, acts with endo-activity, cleaving internal β-1,4 glycosyl linkages.[2] Sequence comparisons suggest that a seventh lytic transglycosylase, named YfhD, may exist in *E. coli* (Swiss Protein accession number P30135) and a similar hypothetical enzyme has been identified in *Haemophilus influenzae* (Swiss-Prot P44587). In addition, the genomes of bacteriophage encode one or more peptidoglycan lytic enzymes to permit the release of the mature phage particles from host cells upon replication. Unlike the enzymes of most phages which act as muramidases, the endolysin from an *E. coli* lambda bacteriophage was also shown to function as a LT.[3]

A search of the National Center for Biotechnology Information (NCBI) databases using the primary sequences of the six characterised LTs of *E. coli,* an Mlt from *Pseudomonas aeruginosa,* and the endolysins of lambda bacteriophage led us to the identify a total of 127 known and hypothetical enzymes from a wide variety of bacteria and bacteriophage.[4] These amino acid sequences were arranged into four families based on alignments, and consensus motifs were identified. LTs were identified in the genomes of both Gram-positive and Gram-negative bacteria, chemoorganotrophs, lithotrophs, phototrophs, and hyperthermophiles. Moreover, the genomes of each of these bacteria, like *E. coli,* possessed more than one type of LT. For example, the *P. aeruginosa* genome encodes homologs of *E. coli* Slt70, YfhD, MltA, and MltD, and four homologs of MltB. This multiplicity and broad distribution suggests that the LTs are ubiquitous in peptidoglycan-producing Bacteria (all but the mycoplasmas) and are thus likely an essential component in the metabolism of peptidoglycan.

3 STRUCTURE-FUNCTION RELATIONSHIP OF LYTIC TRANSGLYCOSYLASE

3.1 Physical Properties

The genes encoding the four hypothetical MltB homologs of *P. aeruginosa* PAO1 genome have been amplified by PCR and cloned as fusion proteins with C-terminal His tags.[5] Expression studies of these genes in *E. coli* in the presence of [^{3}H]palmitate resulted in the labelling of only a single, 40 kDa protein. This labelled protein, like *E. coli* MltB, is thus a lipoprotein and was also named MltB. To facilitate further study, the *P. aeruginosa* MltB gene was engineered to produce an enzyme derivative lacking the N-terminal 17 amino acid residues which included the Cys of the predicted lipidation signal sequence Cys-Ser-Ser. Slow expression of this gene in *E. coli* led to the production of a soluble enzyme derivative which was purified to apparent homogeneity by a combination of affinity, cation-exchange, and gel permeation chromatographies.

The other three non-labelled enzymes were overexpressed in *E. coli* and each was recovered in soluble form and hence named SltB1, SltB2, and SltB3, respectively. Indeed, each of their genes appears to lack the hypothetical lipidation signal sequence and based on amino acid sequence comparisons, these genetically-encoded soluble forms of MltB appear to be rare among the genomes so far sequenced, being confined to only *Vibrio cholerae* and *Mesorhizobium loti,* in addition to *P. aeruginosa.* Each of the three *P. aeruginosa* SltBs have been purified to homogeneity using the same protocol as adopted for MltB.

3.2 Product Distribution and Kinetics of Reactions

The soluble muropeptides released from insoluble peptidoglycan after treatment with MltB and SltB1 have been analyzed by high pH anion-exchange chromatography (HPAEC) with pulsed-amperometric detection.[5] With each reaction, three major and at least seven minor reaction products were resolved by this method (Figure 3). The three major muropeptides were recovered and identified by ESI-MS as di- or tetra-saccharides of GlcNAc and anhMurNAc with attached stem peptides of different compositions. Thus, while the enzymes are presumably localized to different regions of the periplasm, they appear to act similarly to provide new sights for peptidoglycan biosynthesis and/or re-shape the sacculus.

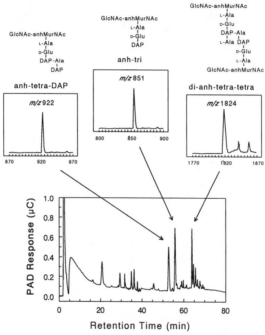

Figure 3   *HPAEC resolution and ESI-MS analysis of muropeptides released from peptidoglycan by* P. aeruginosa *SltB*

Whereas the HPAEC analysis provides excellent resolution of soluble muropeptide products released by LTs, this assay does not readily quantify all reaction products and thus is not amenable for kinetic studies. Such analyses would be further complicated by the use of an insoluble substrate acted upon by a normally membrane-bound enzyme. For these reasons, we have tested a variety of soluble peptidoglycan analogues as substrates for the LTs. However, unlike HEWL and other muramidases, the LTs appear to have a strict requirement for both the *N*-acetyl groups at C-2 and the lactyl-peptide side chain at C-3 of muramoyl residues. Thus, both MltB and SltB are incapable of cleaving soluble chitooligosaccharides (DP 2-6), or their fluorophore derivatives, such as methylumbelliferyl-chitobiose. To further complicate this issue, there is no reaction known to specifically detect the 1,6-anhydromuramoyl products without hydrolysing the β-1,4 linkages between the successive aminosugars.

Given this situation, we have been forced to continue using insoluble peptidoglycan as substrate and devise an alternative convenient assay to process a number samples. This has been accomplished by recovering soluble reaction products released over time and subsequently assaying them for aminosugar content.[6] Using this kinetic assay, the first Michaelis Menten parameters of K_M and k_{cat} for any LT have been determined for *P. aeruginosa* MltB and SltB1.[5] With purified peptidoglycan from *P. aeruginosa* at pH 5.8 as substrate in the presence of 0.1% Triton, the apparent affinity of MltB was found to be approx. 2.5 times greater than that of SltB1, as reflected by their respective K_M values of 72 and 180 μM. However, with a k_{cat} of 2.5 s^{-1}, the overall catalytic efficiency (k_{cat}/K_M) of the latter enzyme is over twice that of the former. Unfortunately, being the first kinetic constants to be established for these enzymes, there is nothing to compare them with. Nonetheless, recognizing that the *in vitro* conditions used to establish these parameters are quite different to those of the natural situation that the enzymes would normally act in, a comparison of the two can be made with reference to differences in their predicted substrate binding sites as predicted by amino acid homologies.

3.3 Three-Dimensional Structures

The three-dimensional structures of the *P. aeruginosa* LTs have not been determined. However, the X-ray crystal structure of a soluble, proteolytic derivative of MltB from *E. coli*, named Slt35, has been solved at 1.7 Å resolution.[7] This protein contains a lysozyme-like fold in its catalytic domain despite no amino acid sequence similarity to any muramidases. Interestingly, the two other LTs for which structural data exist (those of the naturally-occurring soluble LT of *E. coli*, Slt70,[8,9] and lambda phage LT[10]) also contain catalytic domains involving lysozyme folds which are not predicted by sequence alignments. Co-crystallization of Slt35 with GlcNAc-MurNAc-L-Ala-D-Glu led to the identification of four sugar-binding subsites within the enzyme and a single catalytic residue, Glu162, poised between the −1 and +1 binding subsites[11] (Figure 4).

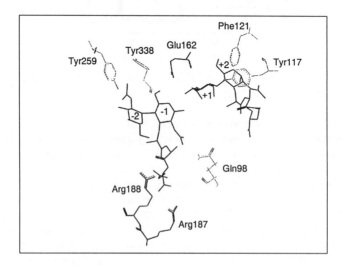

Figure 4   *Three-dimensional structure of* E. coli *Slt35 with bound ligands (PDB: 1D0K)*

In addition to the identification of a single catalytic residue in the putative substrate binding cleft, the crystal structure Slt35 also revealed a number of other contacts between enzyme residues and ligands. Of particular note are interactions between residues and the stem peptide of the ligand at subsite -1. These include hydrogen bonding between Arg188 and the carbonyl of the lactyl moiety and a salt bridge between the free α-carboxyl group of the γ-glutamyl residue of the stem peptide and Arg187 (Figure 4). Arg188 represents a totally conserved residue in the family 3 LTs[4] (Figure 5), suggesting that it plays a key role in positioning substrate for subsequent cleavage. Like the *E. coli* enzyme, many of the family 3 LTs also possess an Arg at position 187. This is the case for *P. aeruginosa* MltB, but not for SltB1 where it is replaced with a proline. Whereas, in general, many factors would contribute to substrate affinity and specificity, we propose that this second Arg at position 187 contributes to the greater apparent affinity determined for SltB1 compared to MltB, as reflected by their respective K_M values. In this context, it is interesting to note that while chitooligosaccharides do not serve as substrate for LTs, they were observed by X-ray crystallography to bind into the active site cleft of the lambda phage enzyme.[10] However, the reducing GlcNAc residue at subsite -1 was angled above the floor of the subsite and did not interact with Glu19, the single acid catalyst of this LT. This further supports the postulate that extra contacts involving the lactyl group of the muramyl residue and/or its associated stem peptide are required to position substrate for catalytic cleavage. To address this issue, we have initiated site-directed mutagenesis experiments to engineer replacements of the two Arg residues at subsite -1 of *P. aeruginosa* MltB. Preliminary data with an Arg188Ala MltB derivative indicates a two-fold reduction of apparent affinity, as reflected by K_M values. These experiments are currently being expanded to involve replacement of Arg187.

```
                               ↓                                    ↓↓
Ec MltB    140 ALNRAWQVYGVPPEIIVGIIGVETRWG  (11) LATLSFNY-----P---RRAEYF
Pa MltB    140 DLQRASRVYGVPPEIIVGIIGVETRWG  (11) LSTLSFSY-----P---RRADFF
Pa SltB1   109 DLARAEKEYGVPAEIIVSIIGVETFFG  (11) LSTLGFDY-----PP---RAEFF
Pa SltB2    99 LLDQVAARYGVDKYTVVAVWGVESDYG  (11) L-TLS-CY--------GRRQSFF
Pa SltB3   150 ILGEVDARYAVDADAVVAIWGMESNYG  (11) LATLA--Y-----E---GRRPEFA
St ORF     138 ALNRAWQVYGVPPEIIVGIIGVETRWG  (11) LATLSFNY-----P---RRAEYF
Kp ORF     136 ALQRAWEVYGVRAIIIVGIIGVETRWG  (11) LATLSFRYP---RPNTSRR-S-W
Yp ORF     139 ALQRAWEVYGVPPEIIVGIIGVETRWG  (11) LATLSFAY-----P---RRATFF
Bb ORF     146 LLNRAAQRYGVPASIIASIIGVETLYG  (11) LATL-FDYP---DPAKPERADMF
Bp ORF     142 LLNRAAQRYGVPASIIASIIGVETLYG  (11) LATLAFDYL---DPAKPERADMF
BppORF     113 VLARAQAEYGVDPATVVAVWGVESNFG  (11) LSTLS-CF--------GRRQSYF
Ng ORF     136 VIDDVAQKYGVPAELIVAIIGIETNYG  (11) LATLGFDY-----P---RRAGFF
Nm ORF     137 LIDDVAQKYGVPAELIVAVIGIETNYG  (11) LATLGFDY-----P---RRRAGF
Tf ORF     162 LLQAVSQKYGVSGPILMGILNIETGFG  (11) NLSLALL------P---GRRR-FF
Cc ORF     131 FLSQIESRYGVPGDILLAVWAMESAFG  (11) MVSLAA------D---GRRR-AW
Vc ORF      98 ELQRIGKQYGVQPRFIVALWGVESN-G  (12) LSTLA--Y----E---GRREEFF
Xf ORF     134 QLKKVEAATGVPAELIVAIIGVESSYG  (11) LYTLAFKYPRSGDPNKLKREVQR
Pm ORF     131 QLEKASQRFGVQKEYLMSLWGMESSFG  ( 8) LSVLATLAF---E---GRRESLF
Ml ORF      97 WLDRIEARFGVDRYILLAIWSMESNYG  (15) LATLG--Y----GDP---KR-SKY
Sp ORF      91 --RTAPGRSLVPPEIIVGIIGVETRWG  (11) LATLSFNY-----P---RRAEYF
Aa ORF     143 QLENASKKFSVPKNYLLALWGMESSFG  (11) LATLAF------D---GRREALF
SpeORF     123 AIAKAASNYQVEPQIIVAIIGEETFYG  (11) LYTLGFYY----EP----RATFF
                .  . .  ..*  ..  ...*. .*               .*            *   .
```

Figure 5 *Amino acid sequence alignment of known and hypothetical family 3 LTs. Hatched and bold residues denote ≥ 80% and 50% sequence identity, respectively; * denotes complete conservation. The line and the arrows identify the "exo-loop" and the putative catalytic Glu and substrate-binding Arg residues, respectively.*

A loop of amino acids was observed to be positioned near the substrate-binding cleft of *E. coli* Slt35.[11] This feature, named the "exo-loop" is believed to block any endo-glycosidase cleavage of peptidoglycan, and thus impose the strict exo-activity of this LT. As depicted in Figure 5, elements of this region (indicated by the line) are maintained in the MltB homologs, including each of the four *P. aeruginosa* enzymes.

4 MECHANISM OF ACTION

To date, there have been no investigations reported which directly address the mechanism of action of the LTs. However, based on both sequence alignments[4] and X-ray crystallography,[7,11] only a single glutamyl is believed to function as a catalytic acid/base. We and others have conducted site-specific replacements to confirm the importance of this residue. Thus, replacement of Glu162 in *P. aeruginosa* MltB with either Asp or Ala led to a complete diminution of detectable activity. Likewise, a Glu162Gln mutant form of *E. coli* Slt35 was inactive.[7] The local environment of this residue is hydrophobic created by the side chains of Ile158, Gln225, Tyr338, and Tyr 344,[7] each of which are highly conserved in the Family 3 enzymes, including *P. aeruginosa* MltB, SltB1, SltB2, and StlB3.[4] This hydrophobic environment would stabilize the protonated state of Glu162 at the optimum pH for its reaction thereby permitting it to function first as an acid to protonate the glycosidic linkage to be cleaved, and then serve as a base to abstract the proton of the C-6 hydroxyl group of the muramoyl residue remaining bound at subsite -1 (Figure 6).

As with the glycosidases, a transition state intermediate involving a oxocarbenium ion is expected to be generated but the LTs do not appear to provide a stabilizing anion/nucleophile. While most glycosidases do indeed provide an anionic residue for such stabilization, the family 20 β-*N*-acetylhexosaminidases, the functionally related family 18 chitinases, and the family 23 goose-type lysozymes also appear to function with a single catalytic residue to catalyze hydrolysis.[12-14] Several lines of evidence suggest that these latter glycosidases invoke substrate-assisted catalysis using the *N*-acetyl group of substrate at subsite -1 to form an oxazoline transition-state intermediate, thereby substituting for a stabilizing anion on the enzyme.[12,13,15] Thus, the LTs could function in a similar manner but instead of catalyzing hydrolysis, the oxazoline intermediate would be attacked intramolecularly by its C-6 oxygen. Early support for this proposal has been provided by inhibition studies using the antibiotic bulgecin. The hydroxyproline residue of this GlcNAc-based competitive inhibitor, which mimics an oxazoline, forms an H-bond to the catalytic Glu478 of Slt70.[16]

5 CONCLUDING REMARKS

Studies to date indicate that the LTs share the same substrate and cleavage site as muramidases, function like the chitinases and related enzymes, but produce a unique 1,6-anhydro-containing product. This latter feature is significant because released fragments of peptidoglycan containing 1,6-anhydromuramic acid have been shown to cause a variety of pathobiological effects. For example, GlcNAc-AnhMurNAc-tetrapeptide, also known as tracheal autotoxin, is released by *Bordatella pertussis*, the etiological agent of whooping cough, and this muropeptide has been shown to induce cell damage in the respiratory tract.[17]

A variety of other apparent effects of muropeptides include pyrogenicity, and the induction of rheumatoid arthritis.[18] Thus, the distinct activity of LTs among peptidoglycan-degrading enzymes makes them an attractive antimicrobial target because a mechanism-based inhibitor should not interfere with innate immunity, *i.e.*, human lysozyme.

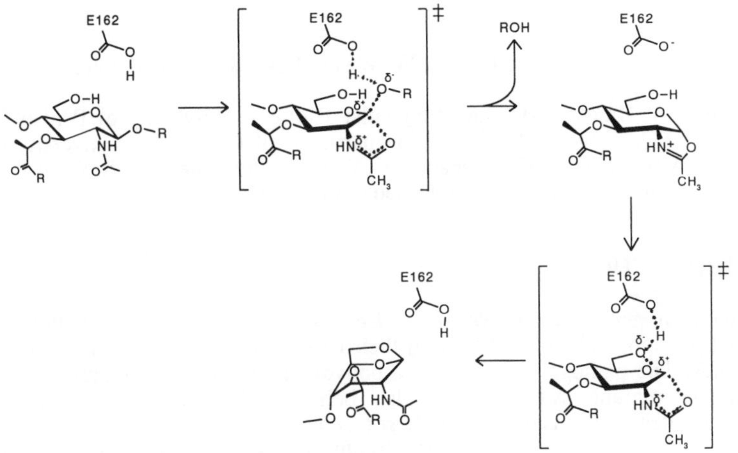

Figure 6 *Proposed mechanism of action of LTs*

References

1. J.-V. Höltje, *Microbiol. Mol. Biol. Rev.*, 1998, **62**, 181.
2. A. R. Kraft, M. F. Templin and J.-V. Höltje, *J. Bacteriol.*, 1998, **180**, 3441.
3. A. Taylor and M. Gorazdowska, *Biochim. Biophys. Acta,* 1974, **342**, 133.
4. N. T. Blackburn and A. J. Clarke, *J. Mol. Evol.*, 2001, **52**, 78.
5. N. T. Blackburn and A. J. Clarke, submitted to *J. Biol. Chem.* 2001.
6. N. T. Blackburn and A. J. Clarke, *Anal. Biochem.*, 2000, **284**, 388.
7. E. J. van Asselt, A. J. Dijkstra, K. H. Kalk, B. Takacs, W. Keck, and B. W. Dijkstra, *Structure,*1999, **7**, 1167.
8. A. M. W. H. Thunnissen, A. J. Dijkstra, K. H. Kalk, H. J. Rozenboom, H. Engel, W. Keck and B. W. Dijkstra, *Nature,* 1994, **367**, 750.
9. E J. van Asselt, A.M.W.H. Thunnissen and B.W. Dijkstra, *J. Mol. Biol.,* 1999, **291**, 877.
10. A. K.-W. Leung, H. S. Duewel, J. F. Honek and A. M. Berghuis, *Biochemistry*, 2001, **40**, 5665.
11. E. J. van Asselt, K. H. Kalk and B. W. Dijkstra, *Biochemistry*, 2000, **39**, 1924.
12. B. L. Mark, D. J. Vocadlo, S. Knapp, B. L. Triggs-Raine, S. G. Withers and M. N. G. James, *J. Biol. Chem.,* 2001, **276**, 10330.
13. I. Tews, A. Perrakis, A. Oppenheim, Z. Dauter, K. S. Wilson and C. E. Vorgias, *Nature Struct. Biol.,* 1996, **3**, 638.
14. L. H. Weaver, M. G. Grutter and B. W. Matthews, *J. Mol. Biol.,* 1995, **245**, 54.
15. S. Knapp, D. Vocadlo, Z. Gao, B. Kirk, J. Lou and S. G. Withers, J. Am. Chem. Soc., 1996, **118**, 6804.
16. A. M. W. H. Thunnissen, H. J. Rozenboom, K. H. Kalk and B. W. Dijkstra, Biochemistry, 1995, **34**, 12729.
17. K. E. Luker, J. L. Collier, E. W. Kolodziej, G. R. Marshall and W. E. Goldman, *Proc. Natl. Acad. Sci. USA,* 1993, **90**, 12652.
18. A. J. Clarke and C. Dupont, *Can. J. Microbiol.*, 1992, **38**, 85.

STRUCTURAL STUDIES OF THE RETAINING GALACTOSYLTRANFERASE LGTC FROM *NEISSERIA MENINGITIDIS*

K. Persson[1], H. Ly[2], M. Dickelmann[3], W. Wakarchuk[3], S. Withers[2] and N. Strynadka[1]

[1]Department of Biochemistry and Molecular Biology, University of British Columbia, Vancouver B.C., Canada, V6T 1Z3, [2]Department of Chemistry, University of British Columbia, Vancouver B.C., Canada, V6T 1Z1, [3]Institute for Biological Sciences, National Research Council, Ottawa, Canada, K1A OR6

1 INTRODUCTION

Bacterial pathogens such as *Neisseria, Helicobacter, Campylobacter* and many more express lipooligosaccharides (LOS) on their cell surface that mimic human cell surface glycolipids. The pathogen uses these carbohydrate structures as a camouflage to evade the human immune system. It is also believed that the bacteria can use these glycoconjugates as means to attach to host receptors.

The synthesis of the bacterial LOS structures is performed by glycosyltransferases, linking saccharides together. These enzymes transfer sugar moieties from nucleotide diphospho-sugars or sugar phosphates to specific acceptor molecules. The enzyme discussed here, α-1,4-galactosyltransferase (LgtC) from *Neisseria meningitidis* transfers a galactose unit from UDP-Gal to a terminal lactose of the LOS structure. The glycosyltransferases have been classified into distinct families depending on sequence similarity and the reaction catalyzed.[1] The glycosyltranferases can also be divided into two major groups, inverting and retaining transferases, depending on the relative anomeric stereochemistries of the substrates and products produced (Scheme 1). Based on analogy with the well-studied glycosidases, inverting glycosyltransferases are believed to follow a direct displacement mechanism with a general base that deprotonates the reactive hydroxyl of the acceptor sugar.[2-4] The retaining transferases are believed to proceed via a double displacement mechanism involving first a nucleophilic attack by the enzyme, resulting in a covalent glycosyl-enzyme intermediate. The intermediate is then displaced by a second nucleophilic attack, by the reactive acceptor sugar hydroxyl, forming a product with the same configuration as the donor sugar.

Until recently, structural information of these glycosyltransferases was very sparse but in the last couple of years several structures of inverting transferases have been solved, giving more insight into this specific catalytic mechanism; β-glucosyltransferase from phage T4[5] SpsA from *Bacillus subtilis*,[6] bovine β-1,4-galactosyltransferase,[7] MurG from *Escherichia coli*[8], the human glucuronyltransferase I[9] and rabbit N-acetylglucosaminyltranferase.[10] Here we present the first crystal structure of a retaining transferase, the α-1,4-galactosyltransferase (LgtC) from *N. meningitidis*. The structure of the enzyme was solved at 2.0 Å resolution in complex with donor and acceptor sugar analogues.[11] The LgtC structure is the first of any nucleotide sugar dependent transferase with structural information on both the donor and acceptor sugars. Recently the second structure of a retaining transferase was published[12], the bovine α-1,3-galactosyltransferse, which, to date, is the transferase structurally most similar to that of LgtC.

2 STRUCTURE OF LGTC

LgtC consists of 311 residues. The 25 C-terminal residues were removed to improve the

a

b

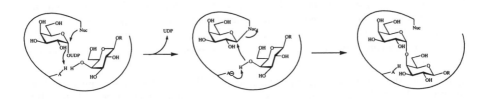

Scheme 1 *The proposed mechanisms of a) an inverting and b) a retaining galactosyltransferase*

solubility and the stability of the enzyme. The C-terminal part of the enzyme is very rich in basic and hydrophobic side chains and is proposed to bind to the negatively charged phospholipids in the bacterial inner membrane via electrostatic and hydrophobic interactions.[13, 14]

The 286 residues in this structure are arranged in two domains. First, a large N-terminal domain composed of a mixed α/β fold that contains the active site. This domain is built up from a seven-stranded β-sheet. The first part of the domain consists of a nucleotide-binding motif with four parallel strands sandwiched between two helices on each side. The rest of the seven-stranded sheet is flanked by four helices on each side. In addition, an antiparallel β-ribbon lies perpendicular to the central β-sheet. The smaller C-terminal domain consists of two helices that are mainly of 3_{10} character. This domain forms a pedestal-like structure that is believed to mediate the membrane attachment (Figure 1).

2.1 Substrate Binding

The donor sugar analogue UDP-2FGal is bound in a deep cleft on top of the central β-sheet. Two long loops, from opposite sides of the binding cleft, fold over the donor sugar analogue as a tight lid. The first loop (residues 75-80) is part of the nucleotide binding motif and the second loop (residues 246-251) is part of a hinge that separates the N- and C-terminal domains. These loops are stabilized by the interactions with the donor sugar as well as by van der Waals interactions between His 78 and Pro 248. The acceptor sugar analogue, 4-deoxylactose is bound in a shallow cleft at the C-terminal end of the α/β-

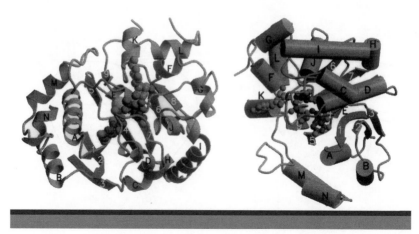

Figure 1 *The overall fold of LgtC. (a) The LgtC structure with bound substrate molecules, depicted in cpk format. The donor analogue is colored dark gray, the acceptor light gray and the manganese ion as a sphere adjacent to the donor. (b) LgtC in an orientation showing the substrate binding N-terminal domain and the membrane attachment C-terminal domain*

domain and the hinge between the two domains. Whereas the donor sugar is almost completely buried by the enzyme, the acceptor sugar is much more accessible to solvent (Figure 2 and 3).

2.1.1 Donor Sugar Analogue Binding. The UDP part of the donor analogue is bound by the nucleotide-binding motif with the uracil base stacking with the conserved Tyr 11. The two phosphates coordinate a manganese ion that is further bound by Asp 103, Asp 105 and His 244. The two aspartic acids are part of a conserved DXD motif that is common among the nucleotide-dependent sugar transferases. The galactosyl moiety of the donor analogue is tucked over the two phosphates so that the plane of the ring is almost parallel to the phosphates. This is an unusual conformation since UDP-Gal and UDP-Glu molecules in the protein data bank generally adopt extended conformations. However, this conformation is also seen in the UDP-Gal bound to the bovine α-1,3-galactosyltransferase[12] suggesting a common mechanism for the retaining transferases. The LgtC galactose sits in a standard full chair conformation and is bound to the enzyme via several hydrogen bonds to conserved active site residues.

2.1.2 Acceptor Sugar Analogue Binding. The acceptor sugar analogue is bound in an open pocket, adjacent to the donor sugar, and is 28% accessible to solvent. The interactions between the enzyme and the acceptor is mainly mediated via van der Waals interactions. Important hydrogen bonds are formed between the acceptor 6'OH and the conserved residues Asp 130 and Gln 189. Additional hydrogen bonds between the acceptor and the enzyme are mediated via water molecules.

2.1.3. Binding of Lactose and the Acceptor Analogue 4-Fluoro-4-deoxylactose. Recently, structures of LgtC and UDP-2FGal in complex with lactose, the real acceptor, was solved, as well as the complex with the acceptor sugar analogue 4-fluoro-4-deoxylactose. These compounds that have chemical groups at their 4'positions (OH for lactose and F for the analogue) bind in identical manner to the previously discussed analogue 4-deoxylactose. However, a chemical group present at this position seems to distort the galactosyl moiety of the donor. Whereas the electron density for the donor galactose is excellent in the donor analogue and donor analogue/4-deoxylactose structures, its electron density in the donor/lactose and donor/4-fluoro-deoxy lactose structures is much weaker. This leads to the assumption that the presence of a 4'-group

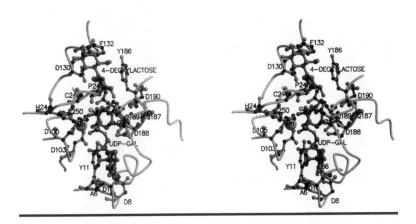

Figure 2 *Stereo view of the active site. The donor sugar analogue, UDP-2FGal (lower molecule) and the acceptor sugar, lactose (upper molecule) are represented as ball-and stick models. Amino acids interacting with the substrates are labeled*

distorts the donor sugar, possibly forcing the galactose to flip between different conformations.

3 REACTION MECHANISM

LgtC follows an ordered bi-bi kinetic mechanism in which the donor UDP-Gal binds first followed by the acceptor lactose. After the reaction has taken place the trisaccharide product is first released and then the UDP moiety. These biochemical data is in perfect agreement with the crystal structure where binding of the donor sugar appears to induce a conformational change (folding of the two loops over the UDP-Gal) which forms the binding pocket for the acceptor sugar. Further evidence for the order of binding is that whereas LgtC containing UDP-2FGal is soluble in the crystallization solution, native protein alone or protein mixed with only lactose will precipitate immediately.

3.1 Possible Nucleophiles

Since LgtC is a retaining glycosyltransferase it was predicted, in analogy with the more studied retaining glycosidases, that the reaction would occur by a double displacement mechanism initiated by a nucleophilic attack on the β-face of the anomeric C″ carbon of the galactosyl moiety to form a glycosyl intermediate. The active site of the enzyme contains a number of conserved aspartic and glutamic acids and it was believed that one of those would function as the nucleophile. However, the crystal structure revealed that there were no residues on the β-face of the galactosyl that could be suitable nucleophiles. The only polar enzyme side chains within 5 Å of the C″ atom are Gln 189 (side chain oxygen) and Asp 153 (side chain nitrogen). Of these, Asp 153 is located 4.2 Å from the anomeric center at a bad angle for a nucleophilic attack but is involved in an intricate hydrogen bonding network, shaping the active site. Gln 189 is located 3.5 Å from the C″ atom in a perfect angle for a nucleophilic attack.[15] It is a common fact that glutamines are poor nucleophiles but a few examples have been described before, e.g. in N- acetyl-

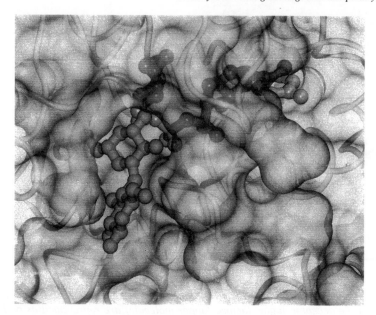

Figure 3 *Transparent surface representation of the binding cleft. UDP-2FGal and 4-deoxylactose are shown in ball-and-stick form*

hexosaminidases where an amide from the substrate itself attacks to form a oxazolinium ion intermediate.[3,4]

3.1.1 Glutamine 189. The Gln 189 hypothesis was investigated by generating the Q189A mutant. This mutant has a k_{cat} value 3% of the wild type enzyme and very similar Km for UDP-Gal although the Km for lactose is seven fold higher than the wild type. The relatively high activity of the mutant enzyme rules out the possibility of Gln 189 being the catalytic nucleophile since mutation of the nucleophiles in the retaining glycosidases usually reduces k_{cat} at least 10^5 fold.[3,4] Also in glycosidases, it has been shown that addition of small molecules to the void created due to the mutation could rescue the catalytic activity. This was also probed in the LgtC Q189A mutant by the addition of formate, acetate, formamide, azide and acetamide, none of which could restore any activity. The conclusion is that the conserved Gln 189 not is the nucleophile but is important for binding the acceptor sugar.

3.1.2 Acceptor Hydroxyl Groups. The other polar atoms located on the β-face of the donor galactose is the 6´OH and 3'OH groups of the acceptor analogue 4-deoxylactose (located at 4.6 Å and 3.2 Å distance respectively). The possibility that the 6´OH performs the first attack on the galactose C″ atom, followed by removal of UDP and a second attack, from the α-face by the acceptor 4´OH producing the desired trisaccharide product was probed in two ways. Firstly, 6-deoxylactose was synthesized and fed to the enzyme. No turnover product was detected which could be consistent with the group being the first nucleophile. However, neither did the 6-deoxylactose function as an inhibitor, a sign of that this compound can not bind to the active site. Secondly, the proposed intermediate galactosyl-β-1,6-lactose was synthesized and added to the reaction mix (enzyme, UDP, manganese and buffer), but no turnover product was observed. These results show that the mechanism where the lactose 6´OH functions as a nucleophile is unlikely. To test if the acceptor 3'OH was involved in the reaction (although at a bad angle relative to the reactive center) the potential intermediate galactosyl-β–1,3-lactose was synthesized. The same experiments as with galactosyl-β-1,6-lactose were performed, with the same negative results.

3.1.3 $S_{N}i$ Mechanism. A third hypothesis is that a covalent intermediate never is formed during the reaction. Instead the enzyme would use a very unusual $S_{N}i$ "internal return" mechanism where a nucleophilic attack by the acceptor 4´ OH and leaving group departure occur at the same time on the same face of the donor sugar. The reaction, that would proceed via a oxocarbenium ion-like transition state, is extremely difficult to prove chemically. However, a similar mechanism has been proposed for glycogen phosphorylase[2] an enzyme that shares some mechanistic features with LgtC. Our recent structure of LgtC with UDP-2FGal and lactose gives some support to this hypothesis since it revealed that the 4' OH in fact is in very close proximity (2.9 Å) to the reactive center of the donor and forms a short hydrogen bond with the glycosidic oxygen of UDP-2FGal.

4 DISCUSSION

The first structures of a retaining galactosyltransferase in complex with both donor and acceptor sugars have been determined. These structures have given further insight into the catalytic mechanism of this class of retaining glycosyltransferases. However, the key residue in the reaction, the nucleophile has not yet been identified even though several of the candidates have been examined by mutagenesis and kinetic studies.

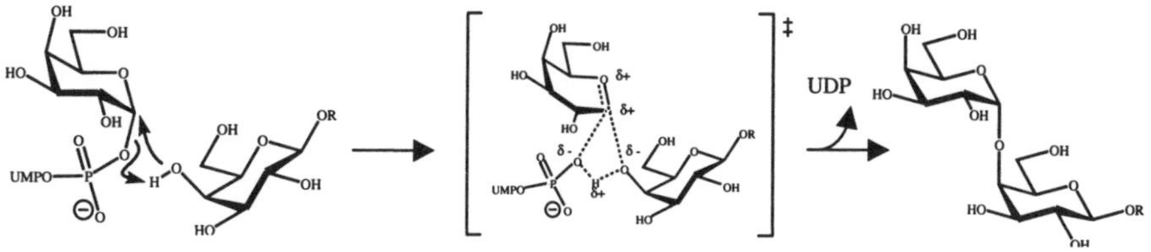

Scheme 2 *Proposed $S_{N}i$ mechanism of LgtC. Nucleophilic attack and leaving departure occur at the same face of the sugar*

References

1. J. A. Campbell, G. J. Davies, V. Bulone and B. Henrissat, *Biochem. J.*, 1997, **329**, 929.
2. M. L. Sinnott, *Chem. Rev.*, 1990, **90**, 1171.
3. G. Davies, S. G. Withers and M. L. Sinnott, in *Comprehensive Biological Catalysis* (Sinnott, M.L., ed) **1** Academic Press, London.1997.
4. D. L. Zechel and S. G. Withers, *Acc. Chem. Res.*, 2000, **33**, 11.
5. A. Vrielink, W. Ruger, H. P. C. Driessen and P. S. Freemont, *EMBO J.*, 1994, **13**, 3413.
6. S. Charnok and G. Davies, *Biochemistry*, 1999, **38**, 6380.
7. L. N. Gastinel, C. Cambillau and Y. Bourne, *EMBO J.*, 1999, **18**, 3546.
8. S. Ha, D. Walker, Y. Shi and S. Walker, *Protein Sci.*, 2000, **9**, 1045.
9. U.M. Ünligil, S. Zhou, S. Yuwaraj, M. Sarkar, H. Schachter and J. Rini, *EMBO J.*, 2000, **19**, 5269.

10. L.C. Pedersen, K. Tsuchida, K. Kitagawa, K. Sugahara, T. Darden and M. Negishi, *Biol. Chem.,* 2000, **275**, 34580.
11. K. Persson, H. Ly, M. Dieckelmann, W. Wakarchuk, S. Withers and N. Strynadka, *Nat. Struct. Biol.,* 2001, **8**, 166.
12. L. N. Gastinel, C. Bignon, A.K. Misra, O. Hindsgaul, J. Sharper and D. H Joziasse, *EMBO J.,* 2001, **20**, 638.
13. W. W. Wakarchuk, A. Cunningham, D. Watson and M. Young, *Prot. Eng.,* 1998, **11**, 295.
14. W. W. Wakarchuk, A. Martin, M. P. Jennings, E. R. Moxon and J. C. Richards, *J. Biol. Chem.,* 1996, **271**,19166.
15. H. B. Burgi, J. B. Dunitz, and E. Shefter, *J. Am. Chem. Soc.,* 1973, **95**, 5065.

AMYLOSUCRASE, A POLYSPECIFIC MEMBER OF FAMILY 13 WITH UNIQUE STRUCTURAL FEATURES

Cécile Albenne, Osman Mirza*, Lars Skov*, Gabrielle Potocki, René-Marc Willemot, Pierre Monsan, Michael Gajhede* and Magali Remaud-Simeon

Centre de Bioingénierie Gilbert Durand, UMR 5504, UMR INRA 792, DGBA INSA, Complexe Scientifique de Rangueil, 31 077 Toulouse Cedex 4 France.
*Protein Structure Group, Department of Chemistry, University of Copenhagen, Universitetsparken 5, DK-2100 Copenhagen, Denmark.

1 INTRODUCTION

Amylosucrase (AS) was first isolated in culture supernatant of the Gram negative bacteria *Neisseria perflava*.[1,2] Other species of the same genus produce AS *ie*: *N. polysaccharea, N. subflava*.[3] Very recently, deduced amino acid sequences sharing 43 % and 48 % of identity with the AS from *N. polysaccharea* have been reported from the genome sequencing of *Deinococcus radiodurans*[4] and *Caulobacter crescentus*.[5]

The first investigations on the extracellular AS from *N. perflava*[1,2] revealed that this remarkable enzyme has the capability of synthesizing an insoluble α-glucan (mainly composed of α-1,4 glucosidic linkages) from sucrose and low amounts of glycogen primer without any participation of nucleotide activated sugars. The enzyme thus appeared very attractive for the synthesis of novel α-glucans and was classified in the category of sucrose glucosyltransferase (EC 2.4.1.4).

To gain new insights in AS mode of action, the AS encoding gene from *N. polysaccharea*[6] was cloned and sequenced and its product was purified to homogeneity. The enzyme consists of a polypeptide chain of 636 amino acids. Opposed to other glucansucrases from lactic bacteria, which belong to glycoside hydrolase family 70,[7,8] AS belongs to glycoside hydrolase family 13. It is the unique glucansucrase that does not contain a circularly permutated $(\beta/\alpha)_8$-barrel.[9] Indeed, sequence alignment and mutational studies revealed that AS possesses a $(\beta/\alpha)_8$-barrel domain and contains the invariant amino acids known as essentials for catalysis in this group of enzymes. By analogy with enzymes of family 13, Glu 328 and Asp286 were proposed to act as the general acid base catalyst and Asp286 as the nucleophile.[6,10]

All these findings suggest that at least for the first part of the reaction (glucosyl-enzyme formation), the catalytic mechanism of AS resembles that of α-amylases. Pure preparations of recombinant native enzyme and mutant enabled new investigations on the catalytic properties[11,12] of the enzyme and successful resolution of the 3D-structures of wild type AS and variant.[13-15] The results of this recent work provide further understanding into the AS unique structural features that control the specificity and the mode of action of this very uncommon enzyme.

2 VARIETY OF REACTIONS CATALYZED BY AMYLOSUCRASE

2.1 Reactions Catalyzed from Sucrose only

The action pattern of amylosucrase on sucrose substrate is very complex.[11] As shown in figure 1, the quantitative analysis of the reaction products reveals that 57 % of the glucosyl residues issued from sucrose are incorporated into an insoluble α-glucan of 9000 Da average molecular mass exclusively composed of α-1,4 linkages. The other glucose moieties coming from sucrose are transferred onto water (4 %), glucose (20 %) or fructose (19 %). Successive transfers onto glucose molecule result in the formation of maltooligosaccharides whereas transfer onto fructose results in the synthesis of two sucrose isoforms: turanose (3-*O*-α-D-glucopyranosyl-D-fructose) and trehalulose (1-*O*-α-D-glucopyranosyl-D-fructose).

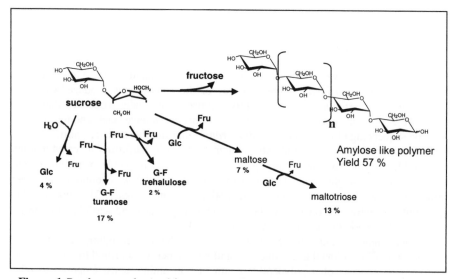

Figure 1 *Products synthesized from sucrose by amylosucrase from N. polysaccharea*
Sucrose concentration 106 mM, product yields correspond to the ratio of glucosyl units incorporated into the product versus the total available glucosyl units of sucrose

The reaction can be separated in two phases. During the first phase, oligosaccharide production is not detectable, hydrolysis and α-glucan synthesis are the major reactions. In addition, the polymer seems to be released only after sufficient elongation suggesting that the α-glucan synthesis proceeds via a processive mechanism. The control of hydrolysis reaction versus polymer synthesis reaction is highly dependant on sucrose concentration (Table 1). Clearly increasing sucrose concentration favors α-glucan synthesis. In addition, kinetic data recorded for sucrose concentrations higher than 20 mM enabled to determine apparent Michaelis-Menten parameters. Values of k_{cat} and K_m are of 1.1 s^{-1} and 26 mM respectively showing that sucrose is acting as a poor substrate.

Table 1 *Ratio of Vi_G (Initial rate of glucose release) to Vi_S (initial rate of sucrose consumption) as a function of sucrose concentration*

Sucrose concentration, mM	6	10	50	106	295
Vi_G/Vi_S, %	73	70	63	55	28

During the second phase, the glucosyl residues are preferentially transferred onto the glucose and fructose moities released during the first phase. When glucose is used as acceptor, the transfer of the glucosyl moiety occurs via the formation of an α-1,4 linkage, the product released in the medium (maltose) can in turn play the role of acceptor and maltotriose is synthesized. In addition, production of turanose and trehalulose indicates that both the 3-hydroxyl and the 1-hydroxyl of fructose react with the glucosyl-enzyme intermediate. However, the fructose acceptor reaction products cannot in turn act as acceptors.

2.2 Reactions in the Presence of Sucrose and Acceptor

To examine the action pattern of AS towards acceptors, reactions in the presence of sucrose and equimolar amount of either glucose, maltose, or maltoheptaose acceptor were carried out.[16] In these conditions, AS catalyses the synthesis of maltooligosaccharides of increasing degree of polymerization (DP). Neither hydrolysis nor polymer synthesis occurred and maltooligosaccharides are released in the medium, the transfer onto acceptors being only limited by the formation of turanose. Addition of acceptors increased k_{cat} 7 to 11-fold. The highest increase is observed for maltoheptaose which is the most efficient acceptor. Amylosucrase is thus a very useful tool for oligosaccharide synthesis.

In the presence of 50 g/L sucrose and 0.1 g/L of glycogen, AS has also been demonstrated to be very efficient in catalysing the elongation of some of the glycogen branchings.[6] Elongation occurs only on a limited number of chains, which are extended from a DP12 to a DP 75. Addition of glycogen greatly reduces the hydrolysis reaction and the formation of oligosaccharides, which is not detectable for glycogen concentration higher than 1 g/L. Moreover, it results in a formidable increase of k_{cat} value, an 80-fold increase being recorded at 10 g/L glycogen. From this observation, it can be concluded that the main biological function of this enzyme is the extension of glycogen-like compound branchings. As revealed by the two-substrate kinetics, AS follows neither an ordered bi-bi nor a ping-pong mechanism.[12] Interestingly, for a given glycogen concentration, initial rate of sucrose consumption increases with sucrose concentration until it reaches a maximum and then decreases for higher concentrations. In addition, the position of the maximum is dependant on the initial glycogen concentration. The reduced activation observed at high sucrose concentration suggests a possible competition between the two substrates.

2.3 Transglycosylation of maltooligosaccharides

Previous studies suggested that AS from *N. polysaccharea* catalyze transglycosylation reactions from maltooligosaccharides.[16] To verify this result and determine the action pattern of the enzyme, maltooligosaccharides of DP ranging from 2 to 6 labeled with pNP at the reducing terminus have separately been tested as substrate.

The reaction products were analyzed at various intervals of time by means of TLC and HPLC analyses. In each of the cases, the decrease of one mole of pNP-labeled maltooligosaccharide of DP_n was correlated to the production of half a mole of pNP-labeled maltooligosaccharide of DP_{n-1} and half a mole of pNP-labeled maltooligosaccharide of DP_{n+1}. This result indicates that AS operates via an exo-acting mode of action. As shown in figure 2, it requires only one glucose unit in the glycon site (-n) for cleavage. In addition, initial rate of consumption of maltooligosaccharides substrate increase with their size and a minimum of four glucose units is required in the aglycon sites (+n) for efficient transglycosylation.

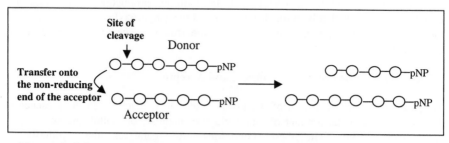

Figure 2 *Schematic representation of the action pattern of amylosucrase onto maltopentaose labeled with pNP at the reducing terminus*

The capability of AS to use sucrose as substrate and to catalyse the transglycosylation of maltooligosaccharides emphasizes the polyspecific character of this enzyme. What are the structural features that control AS polyspecificity and mode of action? Comparison of AS novel structure to other structures of family 13 enzymes and analyses of Glu328Gln mutant in complex with sucrose might help to address these questions.

3 THREE-DIMENSIONAL STRUCTURE OF AMYLOSUCRASE

3.1 Overall Fold

The crystal structure of amylosucrase with TRIS bound in the active site was recently determined to 1.4 Å resolution by Skov et al.[14] It reveals that the monomeric enzyme is organized in five domains named N, A, B, B' and C (Figure 3).

Domain N (residues 1 to 90) shows no structural similarity to any known protein structure. It contains six amphiphilic helices, two of them interacting with two helices (h3 and h4) from the A domain.

Domain A (residues 98-184, 261-395 and 461-550) is the central domain. It contains the characteristic $(\beta/\alpha)_8$-barrel catalytic domain with two protruding loops (loops 3 and 7) which form two separate domains termed B and B'.

Domain B (residues 185-260) between β-strand 3 and α-helix 3 contains two short antiparallel β-sheets. A B-domain is present in most of the members of family 13.

Domain B' (residues 395-460) between β-strand 7 and α-helix 7 is a unique feature for amylosucrase. It is formed by two helices followed by a short β-sheet and is terminated by a short α-helix. It starts right after highly conserved residues in the family and is thought to play an important role in the enzyme specificity (see below).

Finally, domain C (residues 555-628) also found in other α-amylases consists of an eight-stranded β-sheet domain at the C-terminal end of the protein.

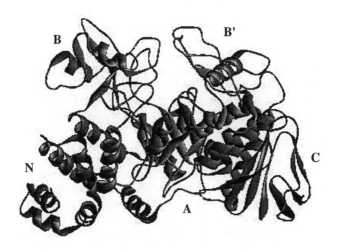

Figure 3 *Schematic representation of the AS structure with the five domains N, A, B, B' and C*

3.2 Common Features with Related Enzymes of Family 13

A structural homology search showed that oligo-1,6-glucosidase[17] (an exo-acting enzyme specific for the cleavage of α-1,6 glucosidic bond) and TAKA-amylase[18] have the highest similarity to AS. A total of 458 Cα atoms of AS and oligo-1,6-glucosidase can be superimposed with a r.m.s. 2.7 Å and 368 superimposable Cα atoms are found in TAKA-amylase:acarbose complex with a r.m.s. of 2.8 Å.

A number of residues of TAKA-amylase:acarbose complex have been identified to play an important role in enzyme substrate interactions around the subsite −1. Interestingly, both Cα and side chains of these conserved residues are found at identical position in AS (Table 2). Particularly the position of the Glu328 and Asp286 supports a similar α-retaining mechanism for AS and α-amylase involving Glu328 as the general catalyst and Asp286 as the nucleophile. This is in agreement with the results obtained from conventional sequence alignment and mutational studies.

3.3 Architecture of the Active Site

Superimposition of the active sites of AS and TAKA-amylase:acarbose complex also revealed major differences in the topology of their active sites.[14] Indeed, AS active site is not located in large cleft like α-amylase active site but is found at the bottom of a pocket. This architecture, usually exhibited by exo-acting enzyme, is mainly due to the presence of the unique domain B' (loop 7), which covers partially the active site entrance.

In addition, subsites corresponding to subsites -2 and -3 of α-amylases are not found in AS. Indeed a salt bridge between Asp144 and Arg509 blocks the bottom of the pocket, leaving space only for the positioning of one glucose moiety in subsite -1. The topography of the active site is thus in total agreement with the action pattern of amylosucrase onto maltooligosaccharide substrates.

3.4 Structural Determinant of Sucrose Specificity

The recent analysis of the active site Glu328Gln AS mutant in complex with sucrose substrate revealed that sucrose binds the active site with the glucosyl ring in a $^4C_1$ conformation at the -1 position and the fructosyl ring in a ^3E conformation at subsite $+1$.[15] The glucosyl ring is maintained by stacking on Tyr147 and Phe250 and by an hydrogen bonding network involving the conserved residues Asp286, His187, Arg284, Asp393, His392 (Table 2) and the residues Asp144 and Arg509 which interact with O4. Only one water molecule associates with the pyranosyl ring at O4.

Interactions between fructosyl moiety and AS involves four direct hydrogen bonds and four water mediated hydrogen bonds. Direct hydrogen bonds involve residues Asp393, Asp394 and Arg446. Interestingly Asp394 and Arg446 belong to the loop 7 and are not conserved in the family. They are thought to primarily determine with Asp144 and Arg509 the specificity for sucrose. The main effect of these interactions is to align the lone pair of the linking O1 for hydrogen bond interaction with Glu328. In addition the distance seen between Asp286 and C1 of pyranosyl ring (3.1Å) is short enough to make a nucleophilic attack possible. Finally, the interactions with the fructosyl ring involving four water mediated hydrogen bonds are in agreement with the fact that fructose must leave easily the active site during the catalysis.

3.5 Reaction Catalysis

In summary, native AS and complex structural analyses confirm that this enzyme acts on the α-glucosidic linkage of sucrose using a retaining mechanism similar to the

Table 2 *Residues found at identical position in native AS and TAKA-amylase:acarbose complex. Both Cα atoms and side-chains are found at identical position*

AS[14]	TAKA amylase:acarbose complex	Interaction with the I-ring of acarbose in TAKA:acarbose complex [18]	Interaction with sucrose in sucrose:Glu328GlnAS mutant[15]
Asp286	Asp206	Hydrogen bond to O6	Hydrogen bond to O6
His187	His 122	Hydrogen bond to O6	Hydrogen bond to O6
Arg284	Arg204	Salt bridge with Oδ1 of Asp206	Hydrogen bond to O2
Glu328	Glu230	Hydrogen bond to N	
Asp393	Asp297	Hydrogen bonds to O2 and O3	Hydrogen bonds to O2 and O3
Tyr147	Tyr82	Stacking of glucose ring	
His392	His296	Hydrogen bond to O of Tyr 82	Stacking of glucose ring Hydrogen bond to O2 and O3

mechanism used by α-amylases. The formation of the glucosyl-enzyme intermediate involves a nucleophilic attack at the C1 of the glucosyl ring (in subsite −1) by Asp286, which is assisted by general acid Glu328 which gives its proton to the O2 of the fructosyl leaving group in subsite +1. However no alternative subsite +1 has been identified in glucose or sucrose complexe and apparently, there is only one access to the active site. Consequently, in a second step, fructose would be released out of the pocket leaving a free access for subsequent attack by the incoming nucleophile.

But, this mechanism leaves a number of questions unresolved. In particular, such a ping-pong mechanism makes a processive elongation of the polymer synthesized from sucrose only impossible. In addition, as shown by kinetic data, biological function of AS is undoubtedly the elongation of the glycogen branchings. The two-substrate steady state kinetics do not follow a ping-pong mechanism. Finally, the control of hydrolysis reaction would be very difficult. For all these reasons, we think that conformational changes induced by the binding of longer oligosaccharides may occur. They could open a new access to the active site or establish an alternative binding site. These changes are not visible in the glucose or sucrose complex but a second weaker glucose binding site was identified at the surface near the active entrance in AS:glucose complex[15]. Further biochemical and structural investigations are now necessary to fully understand the mode of action of amylosucrase.

Ackowledgements - This work was supported by the EU biotechnology project Alpha-Glucan-Active Designer Enzymes (AGADE, BIO4-CT98-0022)

References

1. E. J. Hehre, D. M. Hamilton and A. S. Carlson, *J. Biol. Chem.*, 1949, **177**, 267.
2. E. J. Hehre and D. M. Hamilton, *J. Biol. Chem.*, 1946, **166**, 77.
3. J. Y. Riou, M. Guibourdenche and M. Y. Popoff, *Ann. Microbiol.(Paris)*, 1983, **134B**, 257.
4. O. White, J. A. Eisen, J. F. Heidelberg, E. K. Hickey, J. D. Peterson, R. J. Dodson, D. H. Haft, M. L.Gwinn, W. C. Nelson, D. L. Richardson, K. S. Moffat, H. Y. Qin, L. X. Jiang, W. Pamphile, M. Crosby, M. Shen, J. J. Vamathevan, P. Lam, L. McDonald, T. Utterback, C. Zalewski, K. S. Makarova, L. Aravind, M. J. Daly, K. W. Minton, R. D. Fleischmann, K. A. Ketchum, K. E. Nelson, S. Salzberg, H. O. Smith, J. C. Venter and C. M. Fraser, *Science*, 1999, **286**, 1571.
5. W. C. Nierman, T. V. Feldblyum, M. T. Laub, I. T. Paulsen, K. E. Nelson, J. Eisen, J. F. Heidelberg, M. R. K. Alley, N. Ohta, J. R. Maddock, I. Potocka, W. W. Nelson, A. Newton, C. Stephens, N. D. Phadke, B. Ely, R. T. DeBoy, R. J. Dodson, A. S. Durkin, M. L. Gwinn, D. H. Haft, J. F. Kolonay, J. Smit, M. B. Craven, H. Khouri, J. Shetty, K. Berry, T. Utterback, K. Tran, A. Wolf, J. Vamathevan, M. Ermolaeva, O. White, S. L. Salzberg, J. C. Venter, L. Shapiro and C. M. Fraser, *Proc. Natl. Acad. Sci. U. S. A.*, 2001, **98**, 4136.
6. G. Potocki de Montalk, M. Remaud-Simeon, R. M.. Willemot, V. Planchot and P. Monsan, *J.Bacteriol.*, 1999, **181**, 375.
7. G. Davies and B. Henrissat, *Structure*,1995, **3**, 853.
8. P. M. Coutinho and B. Henrissat (1999) Http://Amfb.Cnrs-Mrs.Fr~Pedro/CAZY.
9. E.A. MacGregor, H. M. Jespersen and B. Svensson, *FEBS Lett.* 1996, **378**, 263.
10. P. Sarcabal, M. Remaud-Simeon, R. M. Willemot, G. Potocki de Montalk, B. Svensson, and P. Monsan, *FEBS Lett.*, 2000, **474**, 33.

11. G. Potocki de Montalk, M. Remaud-Simeon, R. M. Willemot, P. Sarcabal, V. Planchot, and P. Monsan, *FEBS Lett.*, 2000, **471**, 219.
12. G. Potocki de Montalk, M. Remaud-Simeon, R. M. Willemot, and P. Monsan, *FEMS Microbiol.Lett.*, 2000, **186**, 103.
13. L. K. Skov, O. Mirza, A. Henriksen, G. Potocki, de Montalk, M. Remaud-Simeon, P. Sarcabal, R. M. Willemot, P. Monsan and M. Gajhede, *Acta Crystallogr.* 2000, **D56**, 203.
14. L. K. Skov, O. Mirza, A. Henriksen, G. Potocki de Montalk, M. Remaud-Simeon, P. Sarçabal, R. M. Willemot, P. Monsan and M. Gajhede, *J. Biol. Chem.*, 2001, in press.
15. O. Mirza, L. K. Skov, M. Remaud-Simeon, G. Potocki de Montalk, C. Albenne, P. Monsan and M. Gajhede, *Biochemistry*, 2001, in press.
16. G. Potocki, Ph. D. Thesis, INSA Toulouse, 1999.
17. K. Watanabe, Y. Hata, H. Kizaki, Y. Katsube, and Y. Suzuki, *J. Mol. Biol.* 1997, **269**, 142.
18. A. M Brzozowski and G. J. Davies, *Biochemistry* 1997, **36**, 10837.

THREE-DIMENSIONAL STRUCTURE OF MALTO-OLIGOSYL TREHALOSE SYNTHASE

M. Kobayashi[a], M. Kubota[b] and Y. Matsuura[a]

[a]Institute for Protein Research, Osaka University, Suita, Osaka 565-0871, Japan
[b]Hayashibara Biochemical Lab., Amaseminami, Okayama 700-0834, Japan

1 INTRODUCTION

Trehalose (α,α-trehalose), a non-reducing disaccharide containing an α,α-1,1-glucosidic linkage, is known to be widely distributed in yeast, fungi, and plants. It has about 45% sweetness of sucrose, and is gaining significant attention as a food additive and ingredient. Several studies on the biosynthesis of trehalose have been reported, however, none of these systems are well-suited for application to mass production. Later on two efficient enzymes which catalyse trehalose biosynthesis have been found from two different bacterial sources, the mesophilic *Arthrobacter* and an archaebacterium the thermophilic *Sulfolobus acidocaldarius*.[1,2] The two-enzyme system consists of maltooligosyl trehalose synthase (MTSase) and maltooligosyl trehalose trehalohydrolase (MTHase), which catalyse the following reactions in a coupled manner:

$$\text{maltooligosaccharide} \quad \leftrightarrow \quad \text{maltooligosyl trehalose}$$
$$\text{MTSase}$$

$$\rightarrow \quad \text{trehalose} + \text{maltooligosaccharide}$$
$$\text{MTHase}$$

The first reaction is an unusual intramolecular transglucosylation to convert the α-1,4 to an α,α-1,1-glucosidic linkage of the reducing end of maltooligosaccharides, and the second is a hydrolytic reaction at the penultimate α-1,4 linkage of the reducing end resulting in the release of trehalose. The three-dimensional structure of the latter enzyme has been reported recently.[3] We have previously reported the crystallization[4] of MTSase, and describe here its three-dimensional structure determined by means of the single crystal X-ray diffraction method.

2 EXPERIMENTAL

2.1 Production and Purification of Intact MTSase

E. coli strain MMH71-18 harboring an MTSase-overexpression plasmid pKST9, was cultured for 24 hours at 37 °C in a medium containing 2 % maltose, 4 % soy peptone, 2 % yeast extract, 0.1 % Na_2HPO_4 and 100 mg/ml ampicillin, pH 7.0. The MTSase protein was extracted from *E. coli* cells by sonication, and purified by the method described previously.[2] The crude enzyme extract was adsorbed onto a DEAE-Toyopearl 650S column (400 ml) equilibrated with 10 mMTris-HCl buffer (pH 8.0), and eluted with a NaCl gradient of 0 to 0.2 M. The peak fractions were assayed for activity towards maltopentaose by the Somogyi-Nelson method. Fractions containing active enzyme were pooled and adjusted to 1 M ammonium sulfate. After adsorption on a butyl-Toyopearl column (100 ml) the enzyme was eluted with an ammonium sulfate gradient of 1 to 0.5 M. Fractions containing active enzyme were assayed as above, pooled, dialyzed against 10 mM Tris-HCl (pH 8.0), applied onto a Q-Sepharose FF column, and eluted with NaCl gradient of 0 to 0.2 M. The peak fractions after the final step gave a single band on SDS-PAGE. The protein solution was also checked for suitability for crystallization by using a dynamic light scattering instrument, DynaPro-801 TC (*Protein Solutions, Inc.*) at 1.5 mg/ml protein concentration in the same buffer at 5 °C, resulting in a monodisperse profile.

2.2 Chemical Modification of MTSase

Since the crystals of the intact MTSase were of insufficient size and diffraction quality, the enzyme was subjected to reductive methylation of lysine residues essentially according to the method of Means and Feeney.[5] Briefly, the protein solution (10 mg/ml) was dialyzed aganst 0.2 M sodium brate pH 8.5. To 1 ml of this solution 30 μl of 1 M formaldehyde was added followed by additions of 1 M sodium borohydride of 6 μl, and 3 μl after 10 min. This modification reaction was repeated 6 times with 20 min intervals, during which the reaction solution was left on ice. Finally agter 30 min of stirring, 6 μl of 1 M sodium borohydride was added to completely reduce the unreacted formaldehyde. The reaction mixture was left with stirring for 1 hr, and 0.5 g of finely ground ammonium sulfate was added to stop the reaction and precipitate the protein. The modified enzyme was recovered by centrifugation, and dissolved in a minimum amount of 5 mM Tris-HCl buffer (pH 7.5) and dialysed against the same buffer for crystallization experiments.

2.3 Crystallization

Diffraction grade crystals of methylated MTSase were obtained by the hanging drop method at 25 °C using a protein concentration of 30 mg/ml in 0.1 M Tris-HCl (pH 8.5) with 0.2 M $MgCl_2$ and PEG2000 as precipitants. The gradual increase of PEG concentration (11 to 13 w/v %) produced good crystals whose cell dimensions differ from those of previously reported.[4] The new crystal form has the cell dimensions of a = 56.5, b = 68.5, c = 93.8 Å, and β =101.5°, with the monoclinic space group $P2_1$. It contains one molecule of MTSase in the asymmetric unit.

2.4 Structure Analysis

2.4.1. Heavy Atom Derivatives. The native crystals were soaked in various solutions containing different concentrations of heavy-atom compounds. The soaked crystals were checked for isomorphism by measuring changes in their cell dimensions and diffraction intensity, and the difference Patterson function was calculated. Two kinds of useful isomorphous derivatives of uranyl and mercury were prepared by soaking crystals in solutions of 5 mM $K_3UO_2F_5$ and 5 mM mercuric acetate, respectively.

2.4.2. Data Collection and Processing. The intensity data collections for the native crystals were carried out on a Weissenberg camera at the BL18B station at the Photon Factory using wave length 1.0 Å at room temperature, and for the two kinds of derivative crystals intensity data were collected on a Rigaku imaging plate diffractometer R-AXIS IIc with Cu-Kα radiation from rotating anode at room temperature.

2.4.3. MIR Phasing. The isomorphous difference Patterson functions were calculated for each derivative. The positions of the four uranyl atoms could be located in the difference Patterson maps., and that of the mercury atom located in a cross-difference Fourier syntheses phased from a single isomorphous uranyl derivative. The refinement of heavy-atom parameters and the phase determination were carried out by the conventional MIR method using the program SHARP and SOLOMON.

2.4.4. Interpretation of the Map and Model Refinement. The initial structure model of MTSase was built against the electron density map at 1.9 Å resolution using the program O. In this initial model, the electron densities of the residues from 130 to 180 were indistinct probably due to high temperature factors. Starting with this model, the refinement was performed with the slow-cooling protocol using program XPLOR, reducing the R-factor to 0.253 between resolutions of 8.0 to 1.9 Å. The model was corrected manually on the computer grahics using program O against 2Fo – Fc and Fo – Fc maps calculated with the coordinates obtaind after X-PLOR processing. During this process, the region of residues from 130 to 180 was getting clear. At this point, the search for bound water was started using program O. Water oxygen atoms were assigned manually against 2Fo - Fc and Fo – Fc electron density maps. During this stage, the R-factor and free-R was reduced to 0.203 and 0.263. At the next step, methylated lysine were assigned manually agAinst 2Fo – Fc and Fo – Fc electron density maps. Ten mono-methylated lysine and nine di-metylated lysine were assigned reducing the R-factor and free-R to 0.197 and 0.256, respectively.

3 RESULTS

3.1 Overall Structure

The overall structure of MTSase (720 residues) is shown in Fig. 1. It consists of three domains, A (residues 1 - 89 and 202 - 653), B (90 - 201), and C (654 – 720). The domain A is the main part of the enzyme containing incomplete $(\beta/\alpha)_8$-barrel structure, and

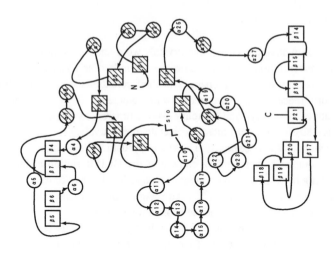

Figure 2   *Topology of the secondary structure arrangement. The elements composing $(\beta/\alpha)_8$-barrel are hatched.*

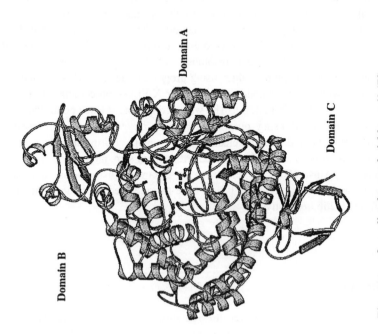

Figure 1   *Overall polypeptide folding of MTSase. Several essential residue side chains at the active site are also shown.*

possesses in several places of the barrel long insertions between the component secondary structure elements. The topology of the secondary structure arrangement is shown in Fig. 2. The $(\beta/\alpha)_8$-barrel is composed of $\beta1$, $\alpha1 + \alpha2$ (kinked), $\beta2$, $\alpha3$, $\beta3$, $\alpha7 + \alpha8$ (kinked), $\beta8$, $\alpha9$, $\beta9$, (lacking 5th helix), S10 (distorted β), $\alpha18$, $\beta12$, $\alpha24$, $\beta13$, and $\alpha26$. The insertions in the domain A occur at the fragments from $\alpha10$ to $\alpha17$ (122 residues) between S10 (6th β) and $\alpha18$ (6th helix), at $\alpha19$ to $\alpha23$ (95 residues) between $\beta12$ (7th β) and $\alpha24$ (7th helix), and at $\alpha25$ between $\beta13$ (7th β) and $\alpha26$ (8th helix). At the C-terminal of the domain A, one short α-helix ($\alpha27$) exists preceding the domain C. The domain B is composed of three α-helices ($\alpha4$, $\alpha5$ and $\alpha6$) and four β-strands ($\beta4$, $\beta5$, $\beta6$ and $\beta7$). The domain B has relatively large average temperature factor compared to other domains, suggesting slightly high domain scale fluctuation. This domain forms the active cleft between the domain A. The domain C, being similar to those in other α-amylases, is composed of anti-parallel β-barrel consisted of eight β-strands $\beta14$ to $\beta21$.

3.2 Active Site

The active site is formed between faces of the C-terminal side of the $(\beta/\alpha)_8$-barrel and the domain B. The cleft is large enough to incorporate the substrate amylose similar to other α-amylases. The catalytic residues Asp228, Glu255 and Asp443 are located at one end of the bottom of the cleft, with the relative three-dimensional arrangement also similar to other α-amylases. The wall of the active cleft is formed by residue fragments of loop 520 - 530, helix around 510, loops 440 – 450 and around 600 in the domain A, and loops 130 - 140, 180 - 190, 190 - 200 and around 90 in the domain B. The bottom of the cleft is formed by loops 45 - 55 and around 600. The cluster of the three catalytic residues is surrounded by the residue fragments, loops around 410, 250 - 260, 440 - 450, helix 270 - 280, loops around 600, around 10, 35 - 40, 85 - 90, 190 - 200, 225 - 230, 345 - 350, 85 - 90, and helix around 390.

The most significant structural features which differs from other α-amylases are that the bottom of the cleft where catalytic residue cluster resides is covered with the helix around 390 and the loop 190 - 200 proximally, and the helix 310 - 320 and the loop 520 - 530 distally. These features are presumably responsible for the intramolecular transglycosylation of this enzyme, enabling enclosure of the once released glucose unit inside the catalytic site, to reform α,α-1,1-glucosidic bond at the reducing end of the cleaved amylose.

3.3 Implications for the Catalytic Mechanism

The enzyme undergoes the intramolecular transglycosylation at the reducing end of the substrate maltooligosaccharide. At the initial stage of the reaction, the reducing end of the substrate is inserted into the active site upon binding onto the active cleft. The structure at the active site is surrounded by the residue fragments forming a pocket suitable for intramolecular reaction, as described in the previous section. The three catalytic residues are located so as to exhibit efficient hydrolysis of the substrate similar to usual α-amylases.[6] However in this enzyme the point of cleavage and simultaneous transglycosylation occurs exactly at the glucoside bond of the reducing end. The spatial

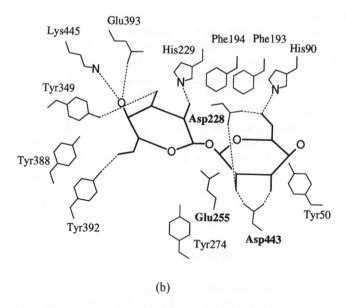

(a)

(b)

Figure 3 *Hypothetical binding mode of carbohydrates in the active site pocket. (a) Binding mode before reaction with an α-1,4-linked maltose. (b) Binding mode after reaction with an α,α,-1,1-linked trehalose.*

extent of the pocket in the three-dimensional structure is a little larger than just to contain one glucose residue. During the transglycosylation reaction process, the once released glucose unit must rotate in some way to form α,α-1,1-glucosidic bond. To enable this movement of a glucose, an extra space is needed at the very end of the pocket, which was demonstrated in the structure. Furthermore, the comparison between α-1,4 and α,α-1,1 bonded structures indicate that the reducing end glucose in the product must have a more tilted configuration than in substrate amylose chain due to axial - axial bond in the α,α-1,1 linkage. This also requires some space in the pocket.

The schematic mode of bindings of substrate and product in the pocket of the active site is shown in Fig. 3. In this figure, hypothetical substrate binding modes derived from the study of molecular modeling using Kendrew-type skeletal wire are shown, comparatively for (a) α-1,4-linked bound maltose, and (b) α,α-1,1-linked bound trehalose. Before the reaction, the inserted substrate may be bound as shown in Fig.3(a). The glucose residue at the position -1 is heavily hydrogen bonded and flanked by several aromatic residues as normally shown in α-amylases.[6] The glucose residue at the +1 site may be hydrogen bonded from the surroundings, His229, Tyr349 and Tyr274. The hydrogen bond between His229 is considered to be important for the recognition of of the substrate, its mutant leading to deactivation. After the reaction, the glucose at +1 site is linked via the α,α-1,1-glucosidic bond with the direction of the ring almost reversed from that before the reaction. The surrounding residues are located favorably to accept the trehalose moiety with possible hydrogen bonds between Tyr349, Glu393, Lys445, Tyr392, and His229. Of these residues, Glu393 and Tyr392 are located on the α-helix α17 which seems to be quite essential in forming the active site pocket. The fact that the mutant Glu393→Ala completely lost the activity suggests the importance of the hydrogen bond between Glu393 in the catalysis. There is a cluster of tyrosines (Tyr349, 388, 392 and 273) in the very end of the pocket, especially the former three being tightly clustered. These tyrosines may also play a leading role in the transglycosylation reaction.

Furthermore, Phe194 is positioned in a similar position of the phenylalanine (or tyrosine) which is considered to help the cyclization in cyclodextrin glucanotransferase (CGTase).[7,8] The same mechanism may be working for the transglycosylation in the present enzyme where α,α-1,1-glucosidic bond formation partly mimics the cyclization of glucose chain.

References

1. T. Nakada, S. Ikegami, H. Chaen, M. Kubota, S. Fukuda, T. Sugimoto, T. Kurimoto, and Y. Tsujisaka, *Biosci. Biotechnol. Biochem.*, 1996, **60**, 267.
2. T. Nakada, S. Ikegami, H. Chaen, H. Mitsuzumi, M. Kubota, S. Fukuda, T. Sugimoto, M. Kurimoto, and Y. Tsujisaka, *Biosci. Biotechnol. Biochem.*, 1996, **60**, 263.
3. M. D. Feese, Y. Kato, T. Tamada, M. Kato, T. Komeda, Y. Miura, M. Hirose, K. Kondo, K. Kobayashi and R. Kuroki, *J. Mol. Biol.*, 2000, **301**, 451.
4. M. Kobayashi, M. Kubata and Y. Matsuura, *Acta Crystallogr.*, 1999, **D55**, 931.
5. G. E. Means and R. E. Feeney, *Biochemistry*, 1968, **7**, 2192.
6. K. Hasegawa, M. Kubota and Y. Matsuura, *Protein Eng.*, 1999, **12**, 819.

7. J. C. M. Uitdehaag, G. J. van Alebeek, B. A. van der Veen, L. Dijkhuizen and B. W. Dijkstra, *Biochemistry*, 2000, **39**, 7772.
8. Y. Matsuura and M. Kubota, in *Enzyme Chemistry and Molecular Biology of Amylases and Related Enzymes*, CRC Press, Tokyo, 1994.

3 Protein Engineering of Carbohydrate-active Enzymes

(GLUCO)AMYLASES, WHAT HAVE WE LEARNED SO FAR ?

B. Svensson[1], J. Sauer[1], H. Mori[1], M. T. Jensen[1], K. S. Bak-Jensen[1], B. Kramhøft[1], N. Juge[2], J. Nøhr[3], L. Greffe[4], T. P. Frandsen[1], M. M. Palcic[5], G. Williamson[2], and H. Driguez[4]

[1]Department of Chemistry, Carlsberg Laboratory, DK-2500 Copenhagen, Denmark, [2]Institute of Food Research, Norwich NR4 7UA, England, [3]Department of Biochemistry and Molecular Biology, University of Southern Denmark, DK-5230 Odense, [4]Centre de Recherche sur les Macromolécules Végétales, C. N. R. S., F-38401 Grenoble, France, [5]Department of Chemistry, University of Alberta, Edmonton, Canada T6G 2G2

1 INTRODUCTION

A lot has been learned since information on α-amylase structure appeared in 1980, but today close to 500 sequences and 50 crystal structures of amylolytic enzymes do not answer all questions on these enzymes. Major issues are i) long-range enzyme-substrate communication, ii) transglycosylation vs. hydrolysis, iii) attack on macromolecular substrates, iv) domain interplay, and v) relations with other proteins. While traditional structure/function relationship investigations developed enormously thanks to advances in structural biology, protein chemistry and recombinant DNA techniques, the emerging era focuses on i) enzymes with custom-made specificity for substrates and proteinaceous inhibitors, ii) new biophysical methods to depict function, and iii) application of genome, proteome, and PIN analysis. This presentation describes the use of mutagenesis and substrate analogs to gain insight into how amylases operate and can be improved.

2 GLUCOAMYLASE (GA)

GA is an inverting enzyme of glycoside hydrolase family 15 (GH15) releasing β-D-glucose from non-reducing ends of various α-1,4/1,6 linked glucosidic substrates.[1,2] It is the first GH reported with a carbohydrate binding module (CBM20) connected to the catalytic domain.[3,4]

2.1 Disaccharide Key-Polar Groups in GA and α-Glucosidase (AG) Hydrolysis

The retaining AGs (GH13 and GH31) share the GA substrates to produce α-D-glucose.[5,6] They differ in distribution and strength of substrate key polar groups critical in transition-state stabilisation.[7] Thus in GA hydrolysis, OH groups of the reducing ring of isomaltose provide 28 kJ×mol^{-1} to binding,[8] compared to $0 - 11$ kJ×mol^{-1} for AGs. The four OH groups at subsite -1 in both enzymes, provide $75 - 90$ kJ×mol^{-1}. While GA recognises the geometry of key polar groups in the R diastereoisomer, shown using conformationally biased C-6-R- and C-6-S-methyl isomaltosides[9], AG selects the S-compound.[7]

2.2 Combination of GA Mutation and Molecular Recognition

GA/acarbose and GA/D-*gluco*-dihydroacarbose structures[10,11] support interpretation of substrate and enzyme engineering, although one interacting pair, Glu180 γ-COO⁻ and maltose OH-2, was identified biochemically prior to crystallography.[12] Comparing activities of wild-type and active site mutants for deoxy and other maltose and isomaltose derivatives, including conformationally biased isomaltosides, allowed correlation of altered GA-substrate binding at one subsite with effects monitored for interactions at another one.[8,13] Thus reduced transition-state stabilisation energy by loss of a sugar-protein H-bond at subsite +1 destabilised sugar OH protein contacts at subsite −1. For isomaltose this loss of transition-state stabilisation at a neighboring ring was overcome by locking the *R*-conformation by C-6-alkylation.[8] Thus weakened sugar-GA contacts troubled GA-induction of substrate conformations suited for transition-state stabilisation at another subsite. Transferred NOE NMR spectroscopy showed GA's conformer selection, the bound conformation of D-*gluco*-dihydroacarbose being the same as in the crystal structure, but different from the solution conformation.[14] Using this technique and modelling of methyl 5'-thio-4-*N*-α- and 5'-thio-4-*S*-α-maltoside GA complexes, superior inhibition by the former stemmed from electrostatic binding between the interglycosidic *N* and the catalyst Glu179, whereas Glu179 turned away from the *S* in the latter analog.[15]

2.3 Probing Domain Cross-Talk in GA

GA was early identified to have a starch binding domain (SBD).[3,4] Today large diversity in CBM structure and specificity is found.[16] GA is thus a prototype of linker-connected multidomain carbohydrases. Simultaneous binding of double-headed inhibitors to both

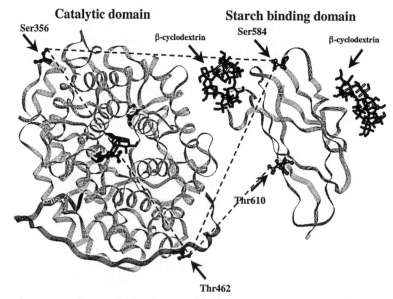

Figure 1 *Positions in the catalytic and starch binding domains of A. niger GA chosen for pairwise cysteine mutation and subsequent fluorophore double labelling*

domains showed them being close together in solution.[17,18] Activity was retained but stability decreased by linker replacement.[19,20] Recently SBD was described to disentangle α-glucan helices in starch granules to promote the enzymatic hydrolysis.[21] The question was then if functionally related interdomain motion occurs. This was addressed by mutational introduction of cysteine (Figure 1) at an exposed position in each domain.[11,22] The thiol groups were then reacted in two steps with fluorescence donor and acceptor molecules to enable effects of ligand binding to be monitored by fluorescence spectroscopy and fluorescence resonance energy transfer (FRET).

Dansyl and fluorescein, the chosen probes, were used before in protein folding studies.[23] GA labelled at S584C in SBD was found by confocal microscopy to adsorb onto and degrade starch granules. Formation of cavities in the granules stopped in the presence of 1 µM acarbose, a potent inhibitor,[24] and fluorescence rested on the granule surface. In contrast, 1 mM β-cyclodextrin, a starch mimic, released GA into the solvent. FRET seemed to occur for a donor and acceptor labelled double mutant S584C-S356C (in the catalytic domain), emitting fluorescence at 515 nm after irradiation at 337 nm. Addition of a bifunctional inhibitor[18] to this derivative, decreased the FRET, indicating that domain-domain contacts were perturbed as a result of bidentate two-domain binding.

2.4 Fusion of GA SBD with Barley α-Amylase 1 (AMY1)

Very few α-amylases possess SBD. It was tried therefore to augment the α-amylase affinity for starch by appending SBD. Barley α-amylase has a 5 β-strand C-terminal domain.[25] As a related maltotetraose-forming amylase sharing this fold has a C-terminal SBD,[26] the homologous SBD from *A. niger* GA was fused to the C-terminus of AMY1. Recently, a fusion of SBD of cyclodextrin glycosyltransferase and *Bacillus* α-amylase degraded raw, but not soluble starch more efficiently than the α-amylase.[27]

The AMY1-SBD gene, also encoding the GA-linker, was expressed in *A. niger* as reported for AMY1.[28] AMY1-SBD, isolated on β-cyclodextrin-Sepharose, migrated as expected in SDS-PAGE, assuming *O*-glycosylation of the linker, and reacted with antibodies against GA and barley α-amylase. The fusion had 2-fold increased activity for soluble starch and starch granules, while activity was the same as of AMY1 for oligosaccharide and amylose DP17. Improved affinity of AMY1-SBD for starch granules compared to AMY1 was indicated by a 10-fold decrease of K_d to 0.13 mg×ml^{-1}.

3 GLYCOSIDE HYDROLASE FAMILY 13 (GH13)

GH13 currently contains 25 hydrolase and tranferase specificities.[29,30] These retaining enzymes comprise a catalytic $(\beta/\alpha)_8$-domain (A), a protruding small domain (B) between β-strand and α-helix 3, and a C-terminal β-sheet domain (C). Several members have extra domains.[31] GH13, GH70 (glucansucrases) and GH77 (amylomaltases) constitute clan GH-H.[32] GH70 has a circularly permuted catalytic $(\beta/\alpha)_8$-barrel[33] and extra domains which are not seen in GH13. Crystal structures are reported of 12 specificities.[29,30,32]

3.1 GH13 Members with Extra N-Terminal Domains

α-Amylases have domains A, B, and C, but are not reported with extra N-terminal domains, found in other family members.[31] Most of such enzymes catalyse reactions on

Table 1 *Examples on GH 13 members with N-terminal domains*

Enzyme	Specificity[a]	Substrate[b]
Isoamylase	h: α-1,6	amylopectin
Cyclodextrinase	h: α-1,4	α-,β-,γ-CDs,starch,pullulan
TVA I	h,tg: α-1,4/α-1,6	pullulan,γ-CD,isopanose,starch
TVA II	h,tg: α-1,4/α-1,6	pullulan,α-,β-,γ-CDs,isopanose,OS
Neopullulanase	h,tg: α-1,4/α-1,6	pullulan,starch
Amylopullulanase	h: α-1,4/α-1,6	pullulan,starch
Pullulanase	h: α-1,6	pullulan,limit dextrins
Limit dextrinase	h: α-1,6	pullulan,limit dextrins
Branching enzyme	tg: α-1,4/α-1,6	amylopectin,amylose
Amylosucrase	tg: α-1,4	sucrose (as donor)

[a]h, hydrolysis; tg, transglycosylation. [b]CD, cyclodextrin; OS, oligosaccharides

α-1,6 bonds or attack α-1,4 near α-1,6 bonds (Table 1). These domains differ in fold and position relative to the $(\beta/\alpha)_8$-barrel. Cyclodextrinase exists in monomer/dimer equilibrium associated to modulation of the activity for β-cyclodextrin and starch.[34]

4 SPECIFICITY AND TRANSGLYCOSYLATION BY GH13 HYDROLASES

Retaining hydrolases (Figure 2) vary in ability to transglycosylate.[30,35] Recently an α-amylase was engineered into a cyclodextrin glucosyltransferase.[36] The diversity of GH13 specificity[29,30] suggests enzymatic synthesis of an array of compounds and α-maltosyl fluoride was used as donor in transglycosylation by AMY1 and limit dextrinase. Catalytic nucleophile D180A/G mutants of AMY1 failed to act as glycosynthase.[35,37]

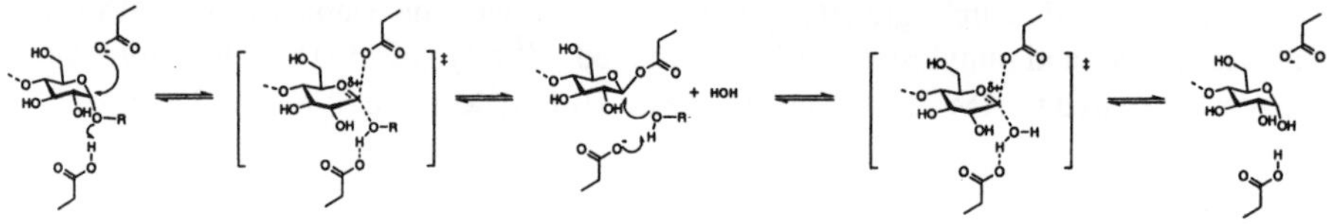

Figure 2 *Retaining mechanism of GH13 hydrolases and transglycosidases*

4.1 AMY1 Branched Oligosaccharide Hydrolysis and Transglycosylation

The action of α-amylases towards branched and linear malto-oligosaccharides reflects the substrate accessibility to specific subsites in the binding cleft. Thus the presence of an α-1,6 branch of an oligosaccharide main-chain modifies both activity and binding mode.

4.1.1 AMY1 Hydrolysis of 6'''-Maltotriose Maltohexaose. Wild-type AMY1 and M298A/N/S mutants (at subsite +1/+2) released glucose from the non-reducing end of the main-chain in a synthetic branched nona-saccharide.[38] The rate of hydrolysis of 76×10^{-4} $s^{-1} \times mM^{-1}$ (0.2 mM substrate) was 15 and 34 times slower than of maltotetraose and maltopentaose. The productive binding mode differed greatly from that of 4-nitrophenyl-maltopentaoside which was cleaved to equal degree at the third and the fourth bond from the non-reducing end. As mutants released glucose with 6-15% activity of AMY1, the shorter side-chains replacing M298 did not advance hydrolysis near the branch.

4.1.2 Transglycosylation by AMY1. α-Maltosyl fluoride was an excellent transglycosylation donor giving high yields with isomaltotriose, isopanose, or cellobiose, modest yields with panose and isomaltose, while sucrose or trehalose were not acceptors.

4.2 Transglycosylation, Substrate and Inhibitor Specificity of Limit Dextrinase

Limit dextrinases are plant debranching enzymes acting on α-1,6 bonds. Isopanose is the minimum substrate for the barley enzyme; activity is much higher on branched malto-oligosaccharides and pullulan consisting of maltotriose units connected by α-1,6 bonds.[39] Transglycosylation with α-maltosyl fluoride as donor and isopanose, panose or maltotriose as acceptor gave high yield of products which are poor substrates for the enzyme. β-Cyclodextrin as acceptor gave low yields, as α-1,6 maltosyl β-cyclodextrin was hydrolysed with $k_{cat} = 4.7$ s^{-1} and $K_m = 60$ μM, comparable to $k_{cat} = 38$ s^{-1} and $K_m = 0.2$ mg×ml^{-1} for pullulan. β-Cyclodextrin, thio-α-1,6 maltosyl- and maltotetraosyl β-cyclodextrins are strong inhibitors of $K_i = 1.4$, 0.46, and 0.56 μM, respectively.

5 ENGINEERING THE SUBSTRATE SPECIFICITY OF AMY1

Detailed insight in enzyme substrate interactions in crystal structures and modelled complexes enabled local manipulation of the surface of the long binding site. In AMY1 kinetically determined subsites comprise −1 through −6 towards the non-reducing end of substrates and +1 through +4 towards the reducing end.[40] The AMY2/acarbose complex covers ligand interactions at subsites −1 through +2,[41] while a modelled AMY2/substrate shows interactions at all ten subsites.[42] Recently the structure was solved of a chimeric *Bacillus* α-amylase/maltodecaose analog complex.[43] Protein engineering in AMY1 include random and biased random mutagenesis of tripeptides in binding β→α loops, saturation mutagenesis of a dipeptide spanning three subsites, and single and dual subsite site-directed mutagenesis (Figure 3).[44-47]

5.1 Local Random and Saturation Mutagenesis

In (β/α)$_8$ enzymes, catalytic and binding residues are at C-termini of β-strands and β→α connecting loops. Sequence alignment and crystal structures of GH13 here identify specificity associated sequence motifs.[29,30] This knowledge was used in attempts to alter specificity by mutation[48,49] and sequences in β→α binding areas were targets in random or saturation mutagenesis coupled with expression screening.[45-47]

5.1.1 β→α 7 Biased Random Mutagenesis. F^{286}VD^{288} in β→α 7 is well conserved in plant α-amylases and varies rather little in GH13 as compared to the succeeding highly variable part of the 7th β→α segment, which in animal α-amylases moves from a position near subsites +1/+2 to interact with substrate at subsite −2.[50] In biased random mutagenesis[46] 5 − 7 new residues were chosen for each of the FVD positions resulting in

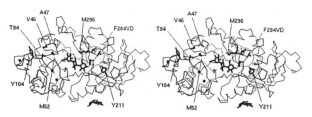

Figure 3 *Stereo view of binding subsite mutations illustrated using AMY2/acarbose.[41] Residues changed by random or site-directed mutagenesis are shown (AMY2 numbering). Acarbose rings fill subsites −1, +1, and +2. A disaccharide unit occupies a surface site*

a theoretical pool of 174 mutant genes. Side-chains causing structural conflicts were excluded, even if common, and a few sequences not reported in α-amylases or in GH13 were allowed. Surprisingly FVG and FGG of the produced five active mutants contained such novel Gly residues. Remarkably these mutants at positions near subsite +1/+2 compensated for low oligosaccharide affinity of the C95A mutant at subsite −6/−5, used to prevent inactivating C95-glutathionylation by the host. These FVD mutants have high activity for starch, but low k_{cat} for oligosaccharides. Gly seems to afford favorable flexibility at the active site although it was not seen naturally. Insight into structure and function is an excellent prerequisite in protein engineering but "random" events are desirable and design of mutant gene pools must consider both wishes when assessing the probability of identifying functional variants by screening. This is relevant in high through-put *in vitro* evolution. The drawback of the FVD mutants is the yield of only 0.1% compared to wild-type and parent AMY1, which possibly stems from perturbed aromatic clusters involving FVD and the secondary structures it bridges. Such problems may be fixed e.g. through *in vitro* evolution using FGG and FVG mutant input genes and screening for variants in which barriers for the protein production are overcome.

5.1.2 Saturation Mutagenesis at the Dipeptide V47S48. In the glycon binding area this dipeptide of β→α 2 spans subsites −5 and −3. V47 in contrast to S48 is very conserved in plants. Saturation mutagenesis by PCR resulted in clones, produced in 9800 *E. coli* colonies, which were used for *Pichia pastoris* transformation and subsequent expression screening. 75 sequenced clones varied considerably at V47. The mutants were ranked with respect to halo size on starch plates and activity in liquid culture for Blue Starch and selected ones were produced on a larger scale. Variation of the relative specificity was achieved, e.g. V47C, V47L-S48A, V47I-S48I have 176, 160, and 122% activity towards Blue Starch; 13, 19, and 8% for DP17 amylose; and 35, 9, and 1.6% for 2-chloro-4-nitrophenyl β-maltoheptaoside compared to wild-type. None of the mutants had activity near that of wild-type for the two soluble substrates.

5.1.3 Designed Mutants of M53 at Subsite −2/−3. GH13 crystal structures possess aromatic substrate binding residues, some of which recur among species, e.g. the stacking Y82 from Y82W83 in Taka-amylase A.[51,52] The barley enzymes have Y52M53 and manipulation of substrate binding was attempted by substituting M53. Examination of the structure suggested that an aromatic residue might block the glycon binding area to disrupt substrate accommodation beyond subsite −3. While M53E has superior activity of wild-type, M53Y/W have 0.9 and 0.1% activity for Blue Starch. Oligosaccharide bond cleavage analysis showed that a barrier indeed had been introduced in these mutants.

5.1.4 The Dual Subsite −6/+4 Mutation. Based on stacking onto substrate by Y104 and Y211 in AMY2/maltodecaose (Figure 3; ref. 42) at the terminal binding subsites −6 and +3/+4, Y105A and T212W/Y were made in AMY1. Comapred to wild-type Y105A has superior activity towards Blue Starch, but lower and very poor activity for DP17 amylose and oligosaccharide, respectively. T212W/Y improved affinity for the latter substrate, and gave normal and decreased activity for amylose and Blue Starch, respectively, compared to wild-type. The double mutant Y105A-T212W had intermediary activity for oligosaccharide and inferior activity for amylose and insoluble starch. These mutants shifted the oligosaccharide binding modes reflecting either the loss or the introduction of aromatic stacking.

5.1.5 The Mechanism of Action in Polysaccharide Hydrolysis. α-Amylases were reported in 1967[53] to cleave high molecular weight amylose by a multiple attack mechanism, the degree of multiple attack (DMA) being defined as the number of bonds hydrolysed after to the first one without dissociation of the enzyme-polysaccharide complex. The double displacement mechanism of retaining enzymes (Figure 2) suggests that in multiple attack the covalent intermediate is hydrolysed followed by a position shift of the product to accommodate glucose residues at and near the reducing end in aglycon binding subsites. DMA was determined for Y105A at subsite –6 to increase to 3.2 compared to 2.0 for wild-type. Presumably stacking at Y105 was disadvantageous for long substrates and its elimination stabilised the enzyme substrate complex. This is in agreement with the Y105A mutant having 140% activity for starch, but extremely low activity for oligosaccharides which seem to depend on the stacking. M298S in contrast has DMA of 1.1. This mutant is interpreted to bind weakly at subsite +1/+2 thus impeding the sliding and formation of productive interactions at the aglycone binding region prior to the next glycoside bond hydrolysis in the multiple attack mechanism.

6 CONCLUSIONS AND FUTURE DIRECTIONS

High resolution structures combined with molecular recognition and tNOE NMR explore enzyme-substrate/inhibitor complexes giving clues to design effectors that exploit the dynamics of the enzyme-substrates interplay. Insight progressed on roles of CBMs and linkers for multidomain polysaccharide metabolising enzymes. Both new combinations with CBMs and use of linkers to join functionalities outside the GH field seem promising.

In GH13 functions of extra domains begin to emerge in addition to the classical SBD (CBM20) present in GH13, GH14, and GH15. To expand this area structures in particular containing longer substrates/products must be solved. Future applications of amylases on renewable resources furthermore motivate improvement in both rational design and *in vitro* evolution of enzymes to achieve useful specificities. Finally, the post-genomic era provides opportunities in data-base mining for interacting proteins to identify e.g. important inhibitors of carbohydrate active enzymes having impact in nutrition and control of physiological processes and defence mechanisms in plants.

Acknowledgements

We thank our colleagues M. S. Motawia, I. Damager, B. L. Møller, K. A. Nielsen, N. Aghajari, R. Haser, X. Robert, T. E. Gottschalk, V. Tran, G. André, E. A. MacGregor, Š. Janeček, J. R. Winkler, M. Harrison, P. Roepstorff, and M. R. Sierks for discussions, materials, and information. The work was supported by the EU Framework Programmes IV (AGADE; BIO4-CT98-0022) and V (GEMINI; QLK1-2000-00811), and the Danish Research Councils' Committee on Biotechnology (grant no. 9502914).

References

1. J. Sauer, B. W. Sigurskjold, U. Christensen, T. P. Frandsen, E. Mirgorodskaya, M. Harrison, P. Roepstorff and B. Svensson, *Biochim. Biophys. Acta*, 2000, **1543**, 275.
2. P. J. Reilly, *Starch/Stärke*, 1999, **51**, 269.
3. B. Svensson, K. Larsen, I. Svendsen and E. Boel, *Carlsberg Res. Commun.*, 1983, **48**, 554.
4. B. Svensson, T. G. Pedersen, I. Svendsen, T. Sakai and M. Ottesen, *Carlsberg Res. Commun.*, 1982, **47**, 55.
5. S. Chiba, *Biosci. Biotechnol. Biochem.*, 1997, **61**, 1233.
6. T. P. Frandsen and B. Svensson, *Plant Mol. Biol.*, 1998, **37**, 1.
7. T. P. Frandsen, Ph.D. Thesis, Odense University, 1997.
8. T. P. Frandsen, B. B. Stoffer, M. M. Palcic, S. Hof and B. Svensson, *J. Mol. Biol.*, 1996, **263**, 79.
9. M. M. Palcic, T. Skrydstrup, K. Bock, N. Le and R. U. Lemieux, *Carbohydr. Res.*, 1993, **250**, 87.
10. A. E. Aleshin, L. M. Firsov and R. B. Honzatko, *J. Biol. Chem.*, 1994, **269**, 15631.
11. B. Stoffer, A. E. Aleshin, L. M. Firsov, B. Svensson and R. B. Honzatko, *FEBS Lett.*, 1995, **358**, 57.
12. M. R. Sierks and B. Svensson, *Prot. Eng.*, 1992, **5**, 185.
13. M. R. Sierks and B. Svensson, *Biochemistry*, 2000, **39**, 8585.
14. T. Weimar, B. O. Petersen, B. Svensson and B. M. Pinto, *Carbohydr. Res.*, 2000, **326**, 50.
15. T. Weimar, B. Stoffer, B. Svensson and B. M. Pinto, *Biochemistry*, 2000, **39**, 300.
16. P. M. Coutinho and B. Henrissat, In Recent Advances in Carbohydrate Bioengineering (Eds. H. J. Gilbert, G. J. Davies, B. Henrissat and B. Svensson, Royal Soc. Chem.), 1999, 3.
17. B. W. Sigurskjold, T. Christensen, N. Payre, S. Cottaz, H. Driguez and B. Svensson, *Biochemistry*, 1998, **37**, 10446.
18. N. Payre, S. Cottaz, C. Boisset, R. Borsali, B. Svensson, B. Henrissat and H. Driguez, *Angew. Chem. Int. Ed.*, 1999, **38**, 974.
19. T. Christensen, B. Svensson, and B. W. Sigurskjold, *Biochemistry*, 1999, **38**, 6300.
20. J. Sauer, T. Christensen, T. P. Frandsen, E. Mirgorodskaya, K. A. McGuire, H. Driguez, P. Roepstorff, B. W. Sigurskjold and B. Svensson, *Biochemistry*, 2001, in press.
21. S. M. Southall, P. J. Simpson, H. J. Gilbert, G. Williamson and M. P. Williamson, *FEBS Lett.*, 1999, **447**, 58.
22. K. Sorimachi, A. J. Jacks, M.-F. Le Gal-Coëffet, G. Williamson, D. B. Archer and M. P. Williamson, *J. Mol. Biol.*, 1996, **259**, 970.
23. M. P. Lillo, J. M. Beechem, B. K. Szpikowska, M. A. Sherman and M. T. Mas, *Biochemistry*, 1997, **36**, 11261.
24. B. W. Sigurskjold, C. R. Berland and B. Svensson, *Biochemistry*, 1994, **33**, 10191.
25. A. Kadziola, J. Abe, B. Svensson and R. Haser, *J. Mol. Biol.*, 1994, **239**, 104.
26. Y. Mezaki, Y. Katsuya, M. Kubota and Y. Matsuura, *Biosci. Biotechnol. Biochem.*, 2001, **65**, 222.
27. K. Ohdan, T. Kuriki, H. Takata, H. Kaneko and S. Okada, *Appl. Environ. Microbiol.*, 2000, **66**, 3058.

28. N. Juge, B. Svensson and G. Williamson, *Appl. Microbiol. Biotechnol.*, 1998, **59**, 385.
29. Š. Janeçek, *Trends Glycosci. Glycotechnol.*, 2000, **12**, 363.
30. E. A. MacGregor, Š. Janeçek and B. Svensson, *Biochim. Biophys. Acta*, 2001, **1546**, 1.
31. H. Jespersen, E. A. MacGregor, M. R. Sierks and B. Svensson, *Biochem. J.*, 1991, **380**, 51.
32. URL: http://afmb.cnrsmrs.fr/~pedro/CAZY/ghf-table.html
33. E. A. MacGregor, H. M. Jespersen, and B. Svensson, *FEBS lett.*, 1996, **378**, 266.
34. K.-H. Park, T.-J. Kim, T.-K. Cheong, J.-W. Kim, B.-H. Oh and B. Svensson, *Biochim. Biophys. Acta*, 2000, **1478**, 165.
35. C. S. Rye and S. G. Withers, *Curr. Opin. Chem. Biol.*, 2001, **4**, 573.
36. L. Beier, A. Svendsen, C. Andersen, T. P. Frandsen, T. V. Borchert and J. R. Cherry, *Prot. Eng.*, 2000, **13**, 509.
37. L. F. Mackenzie, Q. Wang, R. A. J. Warren and S. G. Withers, *J. Am. Chem. Soc.*, 1998, **120**, 5583.
38. I. Damager, C. E. Olsen, M. S. Motawia and B. L. Møller, *Carbohydr. Res.*, 1999, **320**, 19.
39. M. Kristensen, V. Planchot, J. Abe and B. Svensson, *Cereal Chem.*, 1998, **75**, 473.
40. E. H. Ajandouz, J. Abe, B. Svensson and G. Marchis-Mouren, *Biochim. Biophys. Acta*, 1992, **1159**, 193.
41. A. Kadziola, M. Søgaard, B. Svensson and R. Haser, *J. Mol. Biol.*, 1998, **278**, 205.
42. G. André, A. Buléon, R. Haser and V. Tran, *Biopolymers*, 1999, **50**, 751.
43. A. M. Brzozowski, D. M. Lawson, J. P. Turkenberg, H. Bisgård-Frantzen, A. Svendsen, T. V. Borchert, Z. Dauter, K. S. Wilson and G. J. Davies, *Biochemistry*, 2000, **39**, 9099.
44. M. Søgaard, A. Kadziola, R. Haser and B. Svensson, *J. Biol. Chem.*, 1993, **268**, 22480.
45. I. Matsui and B. Svensson, *J. Biol. Chem.*, 1997, **272**, 22456.
46. B. Svensson, K. S. Bak-Jensen, M. T. Jensen, J. Sauer, T. E. Gottschalk and K. W. Rodenburg, *J. Appl. Glycosci.*, 1999, **46**, 51.
47. B. Svensson, K. S. Bak-Jensen, H. Mori, J. Sauer, M. T. Jensen, B. Kramhøft, T. E. Gottschalk, T. Christensen, B. W. Sigurskjold, N. Aghajari, R. Haser, N. Payre, S. Cottaz and H. Driguez, In Recent Advances in Carbohydrate Bioengineering (Eds. H. J. Gilbert, G. J. Davies, B. Henrissat and B. Svensson, Royal Soc. Chem.), 1999, 272.
48. T. Kuriki, H. Kaneko, M. Yanase, H. Takata, J. Shimada, S. Handa, T. Takeda, H. Umeyama and S. Okada, *J. Biol. Chem.*, 1996, **271**, 17321.
49. V. Monchois, M. Vignon, P.-C. Escalier, B. Svensson and R. R. B. Russell, *Eur. J. Biochem.*, 2000, **267**, 4127.
50. M. Qian, R. Haser, G. Buisson, E. Duée and F. Payan, *Biochemistry*, 1994, **33**, 6284.
51. Y. Matsuura, M. Kusunoki, W. Harada and M. Kakudo, *J. Biochem. (Tokyo)*, 1984, **95**, 697.
52. A. M. Brzozowski and G. J. Davies, *Biochemistry*, 1997, **36**, 10837.
53. J. F. Robyt and D. French, *Arch. Biophys. Biochem.*, 1967, **122**, 8.

INCREASING THE THERMAL STABILITY AND CATALYTIC ACTIVITY OF *ASPERGILLUS NIGER* GLUCOAMYLASE BY COMBINING SITE SPECIFIC MUTATIONS AND DIRECTED EVOLUTION

T. P. Frandsen, A. Svendsen, H. Pedersen, J. Vind, and B. R. Nielsen

Novozymes A/S,
Krogshoejvej 36,
DK-2880 Bagsvaerd, Denmark

1 INTRODUCTION

Glucoamylase (GA) is an exo-acting glucanase catalysing the cleavage of α-1,4-linkages with inversion of the anomeric configuration. GA´s from various sources constitute glucoside hydrolase family 15[1] that currently contains above 20 members. GA has been a target for structure-function studies addressing substrate specificity, thermostability, and mechanism of an inverting glucoside hydrolase (for reviews see[2-4]). The *Aspergillus niger* enzyme has thus been extensively investigated by studies focusing mainly on pre-steady-state binding kinetics, site-directed mutagenesis elucidating functional roles of invariant residues, ligand binding thermodynamics, and substrate molecular recognition of wild-type and engineered variants.[2-4]

GA is an industrially extremely important enzyme, used in the enzymatic conversion of starch into high glucose and fructose syrups.[2] Industrial saccahrification are currently performed at 60 °C, while GA's from most sources are unstable at higher temperatures. Development of a thermostable GA, capable of performing industrial saccharification at elevated temperatures, would thus be of significant importance to the starch processing industry. Small achievements towards a thermostable GA were fulfilled through protein engineering of the enzymes from *Aspergillus niger* and *Aspergillus awamori*.[2,6] Several approaches, including replacement of glycines in α-helices, elimination of fragile Asp-X bonds, substitution of asparagine in Asn-Gly sequences, and engineering of additional disulfide bonds into the molecule have been pursued using site-directed mutagenesis.[2,6]

In the present study, we have improved the thermal stability of *Aspergillus niger* glucoamylase by 10 °C using directed evolution. In addition, the thermostabilised variant maintains wild-type catalytic activity at low temperature and the glucose yield under conditions simulating industrial saccharification was at the level of the wild-type enzyme. Finally, variants with increased catalytic activity were developed using site directed mutagenesis.

2 INCREASING THE THERMAL STABILITY OF *A. NIGER* GLUCOAMYLASE

To engineer higher thermostability into *Aspergillus niger* glucoamylase we have combined three approaches; analysis of the three-dimensional structure of a closely related glucoamylase, molecular dynamic modelling simulating conditions at enhanced temperature and using sequence homology to thermostable glucoamylases. Identification of potentially labile or stable substitutions was performed by aligning the *Aspergillus niger* sequence with thermostable glucoamylases from *Talaromyces emersonii*[7], *Thermoascus crustacus* (unpublished) and *Clostridium sp. G0005*.[8] This alignment combined with structural analysis and molecular modelling identified three regions that were subsequently selected for random mutagenesis.

Constructed libraries were screened for variants with increased thermostability using a filter plate assay. Variants with improved stability were subsequently selected, expressed, and purified for evaluation of thermostability, catalytic activity, and under conditions simulating industrial saccharification. The measured half-lifes of selected positive variants at 70 °C and pH 4.5 are shown in table 1. Only minor improvements in the thermal stability from these first generation mutants were obtained. Selected substitutions were subsequently shuffled and the best variants showed an up to a 11-fold increased half-life at 70 °C (Table 1).

Enzymes	Mutations	Half-life (min)
Wild-type		7
1. generation		
GA 1	S119P	9
GA 2	E408R, A425T, S465P, T495A	9
GA 3	A393R	18
GA 4	S386N, E408R	13
GA 5	T2Q, L66V, S394P, Y402F	11
GA 6	T2H	10
GA 7	A11E, E408R	13
GA 8	T2M, N9A, T390R, D406N, L410R	11
GA 9	N313G, F318Y	9
GA 10	S340G, D357S, T360V, S368P	7
GA 11	A1V, L66R, Y402F	12
GA 12	T2Q, A11P, S394R	11
Shuffling		
GA 13	V59A, A393R, T490A	27
GA 14	S56A, V59A, N313G, S356G, A393R, S394R, Y402F	40
GA 15	T2H, A11P, V59A, S119P, N313G, S340G, S356G, E408R, N427M	43
GA 16	N9A, V59A, S119P, A246T, N313G, E342T, A393R, S394R, Y402F, E408R	65
GA 17	N9A, S56A, V59A, S119P, A246T, N313G, E342T, A393R, S394R, Y402F, E408R	68
GA 18	T2H, A11E, V59A, L66R, S119P, N313G, S340G, D357S, A393R, S394R, Y402F, E408R	80
GA 19	S119P, T312Q, Y402F, T416H	70

Table 1 *Half-life of Aspergillus niger glucoamylase and selected improved variants measured at 70 °C and pH 4.5.*

The candidate with the highest thermostability was GA 18 containing 12 substitutions (Thr2His, Ala11Glu, Val59Ala, Leu66Arg, Ser119Pro, Asn313Glu, Ser340Gly, Asp357Ser, Ala393Arg, Ser394Arg, Tyr402Phe, Glu408Arg) compared to the wild-type enzyme. GA 18 was further characterized with respect to stability and catalytic properties. The improvement in thermostability, measured as a T_m-value, of GA 18 compared to the wild-type enzyme was 10 °C (Figure 1). From evaluation of the three-dimensional structure of a closely related glucoamylase from *Aspergillus awamori* var. *X100* feasible explanations for the increased thermostability of GA18 can be suggested. Molecular dynamic modelling thus showed that the N-terminal region of glucoamylase is flexible and may cause decreased thermostability. In GA18 two inter-domain contacts may stabilise the N-terminal region. Thr2 is thus exchanged in GA18 with a His that may form an inter-domain contact to Glu389. In addition, a possible salt-bridge between Ala11Glu and Leu66Arg may stabilize the variant compared to the wild-type enzyme. Other substitutions like Val59Ala, Ser340Glu, and Asp357Ser appears less obvious as they are located at the surface of the enzyme.

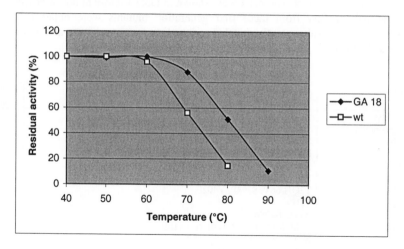

Figure 1 *Temperature stability of GA 18 and the wt enzyme at pH 4.5 measured after incubation of the enzymes 5 minutes at increasing temperatures. Residual activities were measured using 1% (w/v) maltose as substrate and liberated glucose was measured with the God-perid (Bohringer Mannheim) assay*

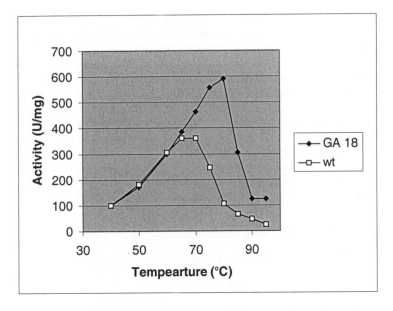

Figure 2 *Activity profile of GA 18 and the wt enzyme as a function of temperature at pH 4.5. Maltose (1 %) was used as substrate*

The specific activity of GA 18 was measured at temperatures ranging from 40 to 95 °C (Figure 2). The specific activity of the mutant was at the level of the wt enzyme even at 40 °C. It has thus been possible to increase the thermal stability by 10 °C and at the same time maintaining the catalytic properties.

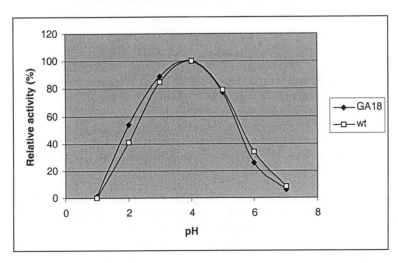

Figure 3 *pH-activity profile of GA 18 compared to the wt enzyme at 40 °C using maltose (1% w/v) as substrate*

Activity optimum of GA 18 was found to 80 compared to 70 °C for the wt enzyme, whereas the pH-activity optimum for GA 18 was identical compared to the wild-type enzyme (Figure 3).

The ability of the mutant to hydrolyse industrial maltodextrin composition to glucose was tested at 60 and 65 °C (Figure 4). At 60 °C, the same maximum glucose yield as for the wild-type enzyme was reached using a 30% lower enzyme dosage of GA 18 compared to the wild-type enzyme. The saccharification evaluation was made on a solution having an initial dry matter concentration of 30% (w/v) maltodextrin. The increased thermal stability thus already shows a positive impact at 60 °C, whereas the effect at 65 °C is more pronounced. Using the wild-type enzyme, it is only able to reach a glucose level of 94.4% compared to 95.8% glucose for the variant (Figure 4). Furthermore the maximum glucose level for the variant is reached after 50 hr whereas the wild-type enzyme still not has reached the maximum after 72 hr.

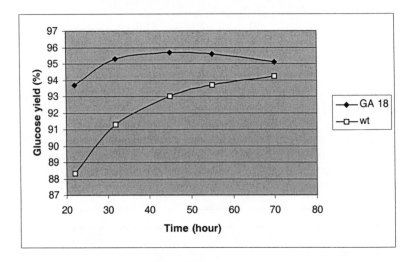

Figure 4 *Saccharification performance of GA 18 and the wild-type enzyme at 65 °C and pH 4.5. Enzyme dosage of the variant and the wt enzyme was 79 µg/g DS. Acid alpha-amylase was added to a dosage of 3 µg/g DS*

3 INCREASING THE CATALYTIC ACTIVITY OF *ASPERGILLUS NIGER* GLUCOAMYLAASE

To increase the catalytic activity of *Aspergillus niger* glucoamylase, a selection of site specific mutations were made based on structural analysis and sequence similarity with *Talaromyces emersonii* glucoamylase showing 3-fold increased rate of catalysis compared to *Aspergillus niger* glucoamylase.[7] In addition, a number of variants from the screening were found to have elevated rate of catalysis. K_{cat}-values for hydrolysis of maltose, isomaltose, and maltoheptaose for a selection of the variants compared to the wild-type glucoamylase are seen in Table 2. The substitutions V111P and G207N both affected the stability in a negative direction with variant 23 and 24 being extremely unstable compared to the wild-type enzyme. In contrast, GA 18 and 19 both showed increased specific activity and thermal stability. Variant 19 exhibited both a 8-fold increased thermal stability and a 4-fold increased catalytic activity on maltose. The ability of the mutant to hydrolyze maltodextrins at industrial relevant conditions (at least 30% dry matter) was, however, quite likely due to product inhibition.

Enzymes	Mutations	$K_{cat}(s^{-1})$		
		Maltose	Maltoheptaose	Isomaltose
Wild-type		11.0	59.7	0.67
GA 18		14.5	-	1.02
GA 19		32.9	-	16.6
GA 20	V111P	17.6	55.5	0.94
GA 21	G207N	17.2	54.0	25.3

Table 2 *Rate of catalysis (k_{cat}) for hydrolysis of maltose, isomaltose, and maltopehtaose by wild-type and GA 18-21*

4 CONCLUSION

Variants of *Aspergillus niger* glucoamylase with both significant increased thermal stability and catalytic activity were developed in the this study. The present approach utilised a combination of site directed substitutions combined with random mutagenesis focusing at a limited number of amino acids in selected regions. Variants with significantly improved thermostability, up to 10 °C increased T_m-value compared to the wild-type enzyme, was thus found in rounds of directed evolution. Furthermore, we have found variants with surprisingly enhanced rate of catalysis on the disaccharides maltose and isomaltose. GA 19 thus showed around 3-fold increased rate of catalysis on maltose and 25-fold increased k_{cat}-value on the α-1,6-linked disaccharide isomaltose.

References

1. P.M. Coutinho and B. Henrissat http://afmb.cnrs-mrs.fr/~pedro/CAZY/db.html
2. C. Ford, *Curr. Opin. Biotechnol.*, 1999, **10**, 353.
3. T.P. Frandsen, H-P., Fierobe and B. Svensson, in *Protein Engineering in Industrial Biotechnology*, ed. L. Alberghina, Harwood Academic Publishers, 2000, Amsterdam, pp 189.
4. J. Sauer, B.W., Sigurskjold, U. Christensen, T.P. Frandsen, E. Mirgorodskaya, M. Harrison, P. Roepstorff and B. Svensson, *Biochim. Biophys. Acta*, 2000, **1543**, 275.
5. B.C. Saha and J.G. Zeikus, *Stärke*, 1989, **41**, 57.
6. H-P.; Fierobe, B.B. Stoffer, T.P., Frandsen and B. Svensson, *Biochemistry*, 1996, **35**, 8698.
7. B.R. Nielsen, J. Lehmbeck and T.P. Frandsen, *Protein Expression and Purification. Submitted*
8. H. Ohnishi, H. Kitamura, T. Minowa, H. Sakai and T. Ohta, *Eur. J. Biochem.*, 1992, **207**, 413.

CYCLODEXTRIN GLYCOSYLTRANSFERASE AS A MODEL ENZYME TO STUDY THE REACTION MECHANISM OF THE α-AMYLASE FAMILY

Joost C.M. Uitdehaag[1*], Lubbert Dijkhuizen[2] and Bauke W. Dijkstra[1]

[1]Centre for Carbohydrate Bioengineering (CCB) and Laboratory for Biophysical Chemistry, University of Groningen, Nijenborgh 4, 9747 AG Groningen, The Netherlands.
[2]CCB and Department of Microbiology, University of Groningen, Kerklaan 30, 9751 NN Haren, The Netherlands
   *current address: Glycobiology Institute, University of Oxford, South Parks Road, OX1 3QU Oxford, United Kingdom

1 INTRODUCTION

The α-amylase family, or glycosyl hydrolase family 13,[1] is an extensively studied enzyme family that comprises many enzymes used in industrial starch-processing. Enzymes from this family have the ability to hydrolyse and synthesize α–glycosidic bonds in poly- or oligosaccharides in various linkage types, such as α(1->1), α(1->4) and α(1->6).[2] Several of them are very suitable catalysts for the synthesis of α-glycosidically linked oligosaccharides. To further improve their applicability it is often desirable to adjust the properties of these enzymes, for instance by protein engineering, which approach is greatly facilitated if the enzyme's interactions with substrates and products are understood in molecular detail. For one member of the α-amylase family, cyclodextrin glycosyltransferase (CGTase), an extraordinary amount of detail about its catalytic mechanism has been elucidated. Here we give a short overview of the X-ray structures of CGTases and complexes with inhibitors, substrates and products, and what we have learnt from them about catalysis and specificity in the α–amylase family.

2 THE REACTION CYCLE OF CGTASE

Like all members of the α-amylase family, CGTase uses a 'double displacement' reaction mechanism.[3] This reaction mechanism is outlined in Figure 1. In the first step, CGTase binds a linear α(1->4) glycosidically linked oligosaccharide, such as present in e.g. its native substrate starch. The enzyme then cleaves the α(1->4) glycosidic bond at the scissile bond position and forms a β(1->4)-linked covalent enzyme-glycosyl intermediate.[4] In the next step, the free sugar (leaving group) departs from the active site to make place for an acceptor group. Subsequently, in the third reaction step, a hydroxyl group of the acceptor group attacks the covalent enzyme-glycosyl intermediate to form a new α(1->4) glycosidic bond in the reaction product. The type of product that is formed depends on the acceptor that is used. For instance, if a water molecule is used as acceptor, this leads to a hydrolysis product. If a sugar is used as acceptor, this leads to a transglycosylation (disproportionation) product. However, CGTase can also use the sugar chain of the covalent intermediate itself as acceptor, which results in a unique circular product, a cyclodextrin (Figure 1). It is this latter activity that has given CGTase its name.

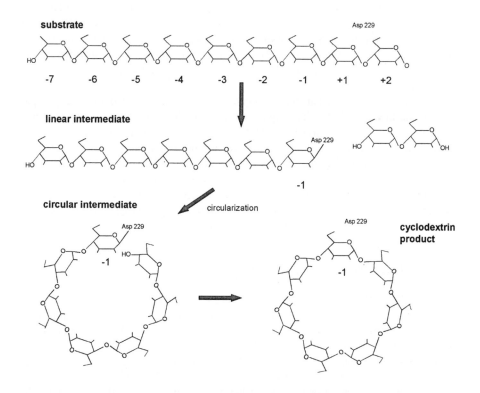

Figure 1 *Stages in the reaction cycle of CGTase. The characteristic reaction path of cyclodextrin formation is shown, although CGTase can use different substrates and acceptors to make a variety of (non-cyclic) products. The location in the enzyme where bond cleavage and synthesis takes place is subsite −1, the catalytic site, where the most important residue is Asp 229 which covalently binds to the intermediate*

3 CGTASE – SUBSTRATE COMPLEXES

The chemical steps of the double displacement mechanism take place in the catalytic site, which is lined with seven residues that are conserved in the α-amylase family.[2] Three of these residues are especially important for catalysis, which in the CGTase of *Bacillus circulans* strain 251 are Asp 229, Asp 328 and Glu 257. In all structurally characterized CGTases[5-10] these residues are found at the bottom of the major, or

"TIM barrel", domain, which is the location that is also observed in other α-amylase family members.[2]

The location of the active site in CGTase was confirmed by X-ray structures of CGTase complexed with maltose,[11] the competitive inhibitor acarbose[12] and maltotetraose substrates.[13] These studies revealed that Asp 229 is suitably oriented to attack the substrate to form an intermediate[12] (Figure 1, see below), and that Glu 257 rearranges after substrate binding to donate a hydrogen that can react with the scissile bond oxygen atom.[12] The role of Asp 328 was revealed by a high resolution complex of CGTase with a maltononaose substrate, which showed that Asp 328, aided by the remaining four conserved catalytic site residues, destabilizes the substrate geometrically and electronically.[4]

Structures of CGTase-substrate complexes have also revealed the many sugar binding subsites in CGTase that help the enzyme to select long oligosaccharide chains from which it can form cyclodextrins.[14] A spectacular structure of CGTase in complex with an acarbose-derived maltononasaccharide inhibitor was the first to show how the enzyme binds glucoside groups at two acceptor subsites (+1 and +2), and seven donor subsites (-1 to -7).[15] This result was the basis for extensive site-directed mutagenesis experiments by many different groups, which succeeded in interfering directly with substrate binding by CGTase, and which altered its reaction-type preferences and cyclodextrin-size specificity (for a review see reference[16]).

4 CGTASE-INTERMEDIATE COMPLEXES

Once bound in the active site, the substrate is cleaved and the donor part of the sugar subsequently binds covalently to Asp 229 to form an intermediate (Figure 1). It is a hot debate in the study of catalysis by glycosyl hydrolases whether Asp 229 binds to the sugar with a full covalent bond, or whether Asp 229 forms a salt-bridge-like interaction with a positively charged oxocarbonium-like sugar moiety.[17] A major step towards resolving this issue was taken by a structure of an E257Q mutant of CGTase in complex with a maltotriosyl oligosaccharide intermediate. In this structure, the stabilizing bond of the intermediate was observed to be covalent.[4]

Comparison of this intermediate structure with a structure of bound substrate shows that the catalytic site of CGTase binds the sugar ring of the intermediate much deeper into its binding site, and that it provides much stronger interactions.[4] Thus, the catalytic site of CGTase binds the intermediate more optimally than it binds the substrate, which might be required to stabilize the intermediate for a sufficient timespan to allow a suitable acceptor to dock. Further proof of this was obtained from the structures of maltose,[11] maltohexaose and maltoheptaose[18] bound to the D229A and E257A/D229A mutants of CGTase. In these complexes, potential van der Waals clashes between the sugar and Asp 229 were removed by the D229A mutation. As a result, the non-reducing end of the sugar chain binds in the catalytic subsite in an intermediate-like position, deep in the catalytic site.[18]

The intermediate stage of the reaction is also the step in which the acceptor binds. To form a cyclodextrin, CGTase must offer the non-reducing end of the intermediate as acceptor, which involves a 23 Å transfer from its 'substrate' position to its 'acceptor' position[15] (Figure 1). This circularization step is probably the rate-limiting step in cyclodextrin formation.[19] A computer model of the reaction path of this unique conformational change suggests that CGTase actively catalyzes the process by transferring the sugar chain along a series of aromatic side-chains in its active site cleft.[20]

5 CGTASE-CYCLODEXTRIN PRODUCT COMPLEXES

Following acceptor binding, the intermediate and acceptor are linked by a new $\alpha(1->4)$-glycosidic bond to form the product. Because all products of CGTase can also be substrates, it was expected that the catalytic site of product-bound CGTase is structurally indistinguishable from that of substrate-bound CGTase. This was largely corroborated by the structure of a CGTase-γ-cyclodextrin product complex and its comparison with the CGTase-maltononaose substrate structure.[21] However, despite both structures being determined under identical experimental conditions, a notable difference was observed in the conformation of the two catalytic site residues Asn 139 and His 140. It was therefore suggested that these residues play a role in an induced fit mechanism that activates the catalytic system after recognition of a long linear substrate, a mechanism which would increase cyclodextrin-forming activity.[18,21]

The structure of the CGTase-γ-cyclodextrin complex also shows how the ring symmetry of the cyclodextrin is distorted in order to fit in the active site.[21] Another structure of CGTase in complex with a thioglucosyl-β-cyclodextrin derivative gives a view of a cyclodextrin that is inhibiting, or has just left, the active site of CGTase, in which case the cyclodextrin ring symmetry is restored again.[22] Both structures reveal that some sugar binding subsites in CGTase are capable of binding sugars in different orientations. Moreover, there exist amino acid residues that are uniquely dedicated to binding either substrate or product, thereby opening the possibility for intervention at specific stages of the reaction cycle through site-directed mutagenesis.[23]

6 CONCLUSIONS

There is no other enzyme in the α-amylase family that has seen such extensive structural studies as CGTase. As a result, the atomic basis of catalytic activity and substrate/product specificity has been unraveled in such detail that CGTase can function as a model for other enzymes in the α-amylase family. Since the catalytic site in the α-amylase family is conserved, the bond cleavage and formation processes seen in CGTase must proceed similarly in the other enzymes of the family. In addition, the studies of CGTase show that the binding of ligands at distant binding sites affects the processes in the catalytic site, leading to induced fit mechanisms that can steer substrate and product specificity. Such mechanisms probably also occur in other enzymes from the α-amylase family.[24] It is possible that those enzymes use a mechanism similar to that of CGTase to convey information from distant subsites into the catalytic site. Only more detailed studies on other members of the α-amylase family will be able to reveal this.

References

1. B. Henrissat and G. Davies, *Curr. Opin. Struct. Biol.,* 1997, **7**, 637.
2. S. Janecek, *Prog. Biophys. Molec. Biol.,* 1997, **67**, 67.
3. D. E. Koshland, *Biol. Rev. Camb. Philos. Soc.,* 1953, **28**, 416.
4. J. C. M. Uitdehaag, R. Mosi, K. H. Kalk, B. A. van der Veen, L. Dijkhuizen, S. G. Withers and B. W. Dijkstra, *Nature Struct. Biol.,* 1999, **6**, 432.
5. B. E. Hofmann, H. Bender and G. E. Schulz, *J. Mol. Biol.,* 1989, **209**, 793.
6. C. Klein and G. E. Schulz, *J. Mol. Biol.,* 1991, **217**, 737.
7. M. Kubota, Y. Matsuura, S. Sakai and Y. Katsube, *Denpun Kagaku,* 1991, **38**, 141.
8. C. L. Lawson, R. van Montfort, B. Strokopytov, H. J. Rozeboom, K. H. Kalk, G. E. de Vries, D. Penninga, L. Dijkhuizen and B. W. Dijkstra, *J. Mol. Biol.,* 1994, **236**, 590.
9. R. M. A. Knegtel, R. D. Wind, H. J. Rozeboom, K. H. Kalk, R. M. Buitelaar, L. Dijkhuizen and B. W. Dijkstra, *J. Mol. Biol.,* 1996, **256**, 611.
10. K. Harata, K. Haga, A. Nakamura, M. Aoyagi and K. Yamane, *Acta Crystallogr.,* 1996, **D52**, 1136.
11. C. Klein, J. Hollender, H. Bender and G. E. Schulz, *Biochemistry,* 1992, **31**, 8740.
12. B. Strokopytov, D. Penninga, H. J. Rozeboom, K. H. Kalk, L. Dijkhuizen and B. W. Dijkstra, *Biochemistry,* 1995, **34**, 2234.
13. R. M. A. Knegtel, B. Strokopytov, D. Penninga, O. G. Faber, H. J. Rozeboom, K. H. Kalk, L. Dijkhuizen and B. W. Dijkstra, *J. Biol. Chem.,* 1995, **270**, 29256.
14. H. Bender, *Carbohydr. Res.,* 1990, **206**, 257.
15. B. Strokopytov, R. M. A. Knegtel, D. Penninga, H. J. Rozeboom, K. H. Kalk, L. Dijkhuizen and B. W. Dijkstra, *Biochemistry,* 1996, **35**, 4241.
16. B. A. van der Veen, J. C. M. Uitdehaag, B. W. Dijkstra and L. Dijkhuizen, *Biochim. Biophys. Acta,* 2000, **1543**, 336.
17. A. J. Kirby, *Nature Struct. Biol.,* 1996, **3**, 107.
18. J. C. M. Uitdehaag, G. J. W. M. van Alebeek, B. A. van der Veen, L. Dijkhuizen and B. W. Dijkstra, *Biochemistry,* 2000, **39**, 7772.
19. B. A. van der Veen, G.-J. W. M. van Alebeek, J. C. M. Uitdehaag, B. W. Dijkstra and L. Dijkhuizen, *Eur. J. Biochem.,* 2000, **267**, 658.
20. J. C. M. Uitdehaag, B. A. van der Veen, L. Dijkhuizen, R. Elber and B. W. Dijkstra, *PROTEINS: Struct. Funct. Genet.,* 2001, **43**, 327.
21. J. C. M. Uitdehaag, K. H. Kalk, B. A. van der Veen, L. Dijkhuizen and B. W. Dijkstra, *J. Biol. Chem.,* 1999, **274**, 34868.
22. A. K. Schmidt, S. Cottaz, H. Driguez and G. E. Schulz, *Biochemistry,* 1998, **37**, 5909.
23. B. A. van der Veen, J. C. M. Uitdehaag, B. W. Dijkstra and L. Dijkhuizen, *Eur. J. Biochem.,* 2000, **267**, 3432.
24. Y. Yoshioka, K. Hasegawa, Y. Matsuura, Y. Katsube and M. Kubota, *J. Mol. Biol.,* 1997, **271**, 619.

4 Domain Structure and Engineering

AN UPDATE ON CARBOHYDRATE BINDING MODULES

Harry J. Gilbert[1], David N. Bolam[1], Lorand Szabo[1], H. Xie[1], Michael P. Williamson[2], Peter J. Simpson[2], Sheelan Jamal[3], Alisdair B. Boraston[4], Doug G. Kilburn[4] and R. Anthony J. Warren[4]

[1]Department of Biological and Nutritional sciences, University of Newcastle upon Tyne, Newcastle upon Tyne NE1 7RU, UK; [2]Department of Molecular Biology and Biotechnology, The Krebs Institute, University of Sheffield, Sheffield S10 2TN, UK; [3]York Strucutral Biology Laboratory, Department of Chemistry, University of York, Heslington, York, UK; [4]Biotechnology Laboratory, Department of Microbiology and immunology, University of British Columbia, Vancouver, Canada V6T 1Z3

1 INTRODUCTION

Polysaccharide composites, exemplified by the plant cell wall, are highly recalcitrant to biological degradation. Enzymes that cleave polysaccharides within complex insoluble structures such as the plant cell wall need to be in intimate and prolonged association with their substrate in order to elicit efficient hydrolysis. The majority of glycoside hydrolases that attack the plant cell wall have a modular structure that comprises catalytic modules and one or more non-catalytic carbohydrate binding modules (CBMs), and it is the CBMs that mediate prolonged contact of the enzyme with its target substrate.[1] This type of modular architecture, however, is not only confined to plant cell wall hydrolases; enzymes such as glucoamylases and chitinases, which degrade the insoluble polysaccharides starch and chitin, respectively, also contain CBMs.[2,3] Although CBMs that bind to crystalline cellulose have been identified in a range of different glycoside hydrolases including cellulases, xylanases, mannanases, xylan esterases, arabinofuranosidases and pectinases, it is becoming increasingly apparent that the ligand specificity of CBMs are often similar to the substrate specificity of the enzyme's catalytic module.[4,5]

At the 3rd Carbohydrate Bioengineering Meeting a new nomenclature for CBMs was proposed, a scheme that has now been generally accepted.[6,7] CBMs have been classified into families based on primary structure similarity. Currently there are more than 27 CBM families.[8] In addition, CBMs have been assigned to three main types, A, B and C.[7] Type A CBMs bind to insoluble crystalline polysaccharides such as cellulose and chitin; Type B modules interact with extended binding sites (*e.g.* 3-6 sugar residues) to single chains of soluble and insoluble polysaccharides, and Type C proteins interact with mono- or disaccharide moieties within polysaccharides. The objective of this report is review advances in the CBM field since the last Carbohydrate Bioengineering Meeting in April 1999.

2 LIGAND SPECIFICITY

The ligand specificity of CBMs in families 1, 5 and 10 are very similar; all the proteins in these families bind to highly crystalline polysaccharides such as bacterial microcrystalline cellulose and chitin.[1] In contrast, the ligand specificity of CBMs from other families can be highly variable. For example, the majority of CBM2 proteins bind to crystalline cellulose,[9] however, some members of this family associate primarily with xylan.[10] This family has thus been subdivided into CBM2a (cellulose binders) and CBM2b (xylan binders). Similarly, CBM4 also contains proteins that exhibit a wide a range of different ligand specificities. CBM4s have been described that bind to single cellulose chains, laminarin (β1,3-glucose polymers) and xylan.[11] Although some CBM22 proteins bind primarily to xylan, several members of this family do not appear to interact with any of the polysaccharides commonly found in plant cell walls.[12] The functional significance of these non-xylan binding CBM22 proteins remains to be established. In general, CBMs bind preferentially to a single ligand, however, a novel protein (CelC) expressed by the anaerobic fungus *Piromyces equi* binds to galactomannan, soluble derivatised cellulose, glucomannan and xylan. In addition, Family 6 CBMs from *Clostridium stercorarium* (F-9) and family 4 CBMs from *Rhodothermus marinus* have been reported to bind tightly to both cellulose and xylan.[13]

It is well established that the ligand binding site of CBMs that bind to crystalline cellulose is a planar surface that contains three aromatic residues. It has been proposed that these modules stack against exposed sugar rings in the cellulose chains.[14] This now seems unlikely as only at the two apical cellulose chains are the sugar rings fully exposed. If these were the CBM binding regions the capacity of the polysaccharide for the protein is estimated to be ~1 μmol/g cellulose, far less than the observed capacity of ~10-15 μmol CBM2a/g cellulose. An alternative mode of binding has recently been proposed by McLean et al.[15] They suggest that the exposed tryptophans form angled or offset parallel stacking interactions with the sugar rings. The exposed tryptophan residues may bind along one cellulose chain or straddle several 'steps' that comprise the cellulose surface. This model predicts that the middle tryptophan and one of the flanking aromatic residues bind to cellulose, and is thus consistent with the observation that replacing the central tryptophan with alanine causes a larger reduction in the affinity of the CBM2a for cellulose than replacing either of the other solvent exposed aromatic residues.[15]

The mobility of CBM2as when bound to cellulose has also been a subject of considerable interest. CBM2as appear to bind irreversibly to crystalline cellulose, however, it was unclear whether the bound protein is mobile on the surface of the ligand. The surface diffusion of CBM2as on cellulose was demonstrated using fluorescence recovery after photobleaching, indicating that CBM2as do in fact move acrosst the cellulose surface. The data showed that diffusion rates varied from 10^{-11} to 10^{-10} cm^2/s, and suggest that diffusion rates on the surface of the substrate do not limit cellulase activity.[16]

Although Type A CBMs interact with crystalline cellulose, it is not clear whether they share a common binding site on the cellulose surface. A recent study suggests that these proteins can in fact interact with different regions of crystalline cellulose, and may provide a rationale for the presence of multiple CBMs from different families in the same enzyme. Carrard et al.[17] showed that when different CBMs, belonging to families 1, 2 and 3a were non-covalently attached to a cellulase, the enzyme was targeted to different sites on the cellulose surface depending on which CBM it was attached to. It would appear that CBM3a binds to the same sites as CBM1 and CBM2a, but also targets the cellulase to additional regions of the cellulose that the other two proteins do not recognize.

3   THERMODYNAMICS OF LIGAND BINDING

The Gibb's free energy and, thus, the association constants for binding of CBMs to their respective ligands is determined by changes in enthalpy and entropy. Sources of enthalpy are polar interactions, such as ionic or hydrogen bonds or charge-charge interactions, and van der Waals interactions. Changes in entropy arise from alterations in the translational and rotational freedom of the polypeptide and carbohydrate, amino acid side chain freedom and solvent (*i.e.* water) re-organization. Precisely how these factors combine to create a thermodynamic profile for a given CBM-carbohydrate interaction is varied. The binding of CBM2as to crystalline cellulose is mediated primarily by entropic forces. The release of water molecules from the surface of the protein increases the entropy of the system; conformational restriction of the ligand on binding (and thus a net decrease in entropy) is not a factor as 'free' crystalline cellulose is already fixed in conformation by the multiple inter- and intrachain hydrogen bonds between chains in the crystal[18]. In contrast, the binding of CBMs to soluble polysaccharides is driven mainly by enthalpic forces. The increase in entropy through the release of ordered water molecules is much less than the reduction in entropy through the conformational restriction of the ligand[19].

CBM22-2   CBM2b-1

Figure 1 *The three dimensional structure with two CBMs, one with a deep cleft (CBM22-2) the other with an exposed binding site (CBM2b-2)*

To further probe the thermodynamic forces driving CBM polysaccharide interactions, the thermodynamics mediating the binding of mutant CBM2bs, which form less hydrogen bonds than the native protein, with its target ligand, xylan, have been assessed. The data, presented in Table 1, show that the affinity of the mutant proteins for the polysaccharide were broadly similar to the unmodified CBM, although the mutations caused a large increase in entropy and reduction in enthalpy (Table 1).[20] It was proposed that the loss of the hydrogen bonds allows the CBM-ligand complex greater motional freedom, which provides an increase in entropy that almost offsets the loss of enthalpy caused by the loss of the hydrogen bonds. The remaining hydrogen bonds may also weaken at the same time, thus allowing further motional freedom, and further shifting the balance between enthalpy and entropy. This redistribution of energies is only possible where the bound ligand has the possibility of increased motion. Therefore, this analysis only applies to cases, exemplified by family 2b CBMs, where the binding site is exposed on the surface (Figure 1). Relatively buried binding sites, such as those found in some lectins and bacterial periplasmic sugar binding proteins, have no possibility of affording motional freedom to the ligand. In such cases the loss of a hydrogen bond can cause a dramatic decrease in overall affinity. This view is supported by the removal of Glu 138 from CBM22-2 of *Clostridium thermocellum* Xyn10B, which forms a hydrogen bond with the target ligand xylan. The E138A mutant exhibits no affinity for xylan, and the three dimensional structure of the mutant protein showed that the mutation caused no perturbation to the ligand binding site.[21] This is consistent with the location of the binding site of CBM22-2, which comprises a relatively deep cleft, thus restricting motional freedom of the bound ligand (Figure 1).

4   CO-OPERATIVITE BINDING BETWEEN CBMS

It is well established that many glycoside hydrolases contain multiple copies of CBMs that belong to the same family. The biological significance of this type of molecular architecture has recently been investigated. Thus, Linder et al. fused two CBM1 proteins

Table 1 *Thermodynamic parameters for the binding of CBM2b-1 wild type and mutants to xylan*

Protein	K_a (x 10^3 M^{-1})	ΔG^o (kcal mol^{-1})	ΔH^o (kcal mol^{-1})	$T\Delta S^o$ (kcal mol^{-1})
WT	6.42 (±0.90)	-5.2 (±0.06)	-9.3 (±0.48)	-4.1 (±0.51)
Q288A	4.40 (±0.64)	-5.0 (±0.05)	-8.7 (±0.22)	-3.7 (±0.58)
E257A	4.45 (±0.11)	-5.0 (±0.04)	-9.0 (±0.23)	-4.0 (±0.52)
N292A	5.53 (±0.27)	-5.1 (±0.08)	-7.4 (±0.44)	-2.3 (±0.29)
E257A/Q288A	3.08 (±0.53)	-4.8 (±0.05)	-7.8 (±0.31)	-3.0 (±0.37)
Q288A/N292A	4.49 (±0.75)	-5.0 (±0.10)	-6.9 (±0.25)	-1.9 (±0.45)
E257A/Q288A/N292A	2.77 (±0.40)	-4.7 (±0.09)	-5.4 (±0.49)	-0.7 (±0.53)

together via a flexible linker sequence. The affinity of the hybrid protein was considerably higher than the original CBMs, suggesting that when linked the CBMs were acting co-operatively to bind cellulose.[22] Recently, Bolam et al. showed that co-operativity can also occur between CBMs *in vivo*. The affinity of *Cellulomonas fimi* Xyn11A for xylan and cellulose is approximately 20 times high than derivatives of the protein that lack either of its two copies of CBM2b.[23] In addition, the two CBM29s in CelC bind co-operatively to a range of polysaccharides that include mannan, cellulose and xylan; and the triplet CBM6 modules from *C. stercorarium* (NCIB 11745) bind cooperatively to cellulose and xylan (Boraston *et al.*, submitted). It appears therefore that cooperativite binding between multiple copies of CBMs from the same family that are present in the same protein is an emerging trend.

5   THREE DIMENSIONAL STRUCTURE OF CBMS

Since the 3rd Carbohydrate Bioengineering Meeting in 1999 the three-dimensional structure of numerous CBMs have been solved by X-ray crystallography and NMR spectroscopy. Thus, the solution structure of CBM10 from *Pseudomonas cellulosa* Xyn10A revealed a β-stranded protein that contains a planar ligand binding surface comprising three aromatic residues. Interestingly this Type A protein has the same fold as oligonucleotide/saccharide fold (OB), however, the ligand binding site of OB proteins and the CBM10 are in different locations (Figure 2).[24] The structures of several Type B CBMs have also been solved. The crystal structure of the xylan-binding CBM22-2 module from Ct Xyn10B revealed a relatively deep cleft, which contains two aromatic residues (Trp 55 and Tyr 103) and three hydrophilic amino acids that are conserved in many CBM22 modules, and are present on the surface of the cleft (Figure 1).[25] Mutagenesis studies showed that these five residues were critical to ligand binding, and NMR titrations with xylohexaose demonstrated that the chemical shift of Trp 55 moved when the protein bound the oligosaccharide.[21] The solution structure of CBM4-2 from *Rhodothermus marinus* Xyn10A has also recently been solved by NMR. The β-stranded protein contains two aromatic residues at the top of a relatively deep cleft. The chemical shifts of these aromatic amino acids changed on titration of the protein with xylohexaose. Interestingly the aromatic residues in Cf Cel9B CBM4-1 were located much further down the binding cleft, which is likely to be linked to the different ligand specificity of these proteins (Cf Cel9B CBM4-1 binds only glucose polymers while Rm Xyn10A CBM4-2 binds to both xylose and glucose polymers).

Three other Type B CBMs have been solved by X-ray crystallography. The CBM15 from Pc Xyn10C is a β-stranded protein with a deep cleft that contains two tryptophan residues which have a perpendicular orientation with respect to each other. The structure of the protein in complex with xylopentaose showed that the two aromatic residues bound to xylose moieties *n* and *n+2*, and the ligand adopted a three-fold helical conformation typical of xylan (Figure 2). A further xylan-binding CBM belonging to family 6 has also recently been solved. The protein contains two deep clefts, one of which is partially covered with a loop. The other cleft contains two aromatic residues that probably interact with xylose residues *n* and *n+1* of the polymeric ligand. Mutation of either of these amino acids abolishes ligand binding, and NMR spectra of CBM6 titrated with xylohexaose showed that the chemical shifts of these two aromatic residues were affected by the addition of the sugar. The third Type B structure to be solved was CBM29-2 of Pe CelC. The protein, which is primarily β-stranded, contains a deep cleft that contains three aromatic residues that are in a twisted orientation with respect to each other. The crystal structure of the protein bound to mannohexaose and cellohexaose has also been elucidated. The ligands adopt a twisted conformation and the interactions between the two oligosaccharides and the CBM were very similar, with the three aromatic residues interacting with sugars *n, n+2* and *n+4*, respectively.

The crystal structures of two Type C CBMs have recently been solved in complex with their target ligands. CBM9-2 from *Thermtoga maritima* Xyn10A binds to the reducing ends of cellulose and xylan chains, disaccharides and, to a lesser extent, monosaccharides[26]. The structure of CBM9 in complex with various β-1,4 polymers of glucose and xylose revealed a binding site that may be described as a blind canyon (Figure 2). The CBM formed numerous hydrogen bonds with the terminal reducing end sugar. The protein also contained two tryptophans that sandwiched the ligand at the terminal glycosidic bond, and appeared to stack against the sugar rings of the reducing end disaccharide.[27] The second Type C CBM whose structure is known is the family 13 CBM appended to a family 10 xylanase from *Streptomyces olivaceoviridis*. The CBM has three similar subdomains, defined as α, β and γ as suggested from a triple-repeat sequence, which are assembled around a pseudo-3-fold axis, forming a galactose-binding lectin fold similar to the ricin B-chain (Figure 2).[28] The affinity of a single binding site on CBM13 for soluble xylan is approximately 12-fold lower than the complete module, indicating that the three sites act co-operatively to bind xylan. Mutagenesis studies showed that xylan could occupy two of the three sites at any given time.[29]

In several CBMs metal ions, principally calcium, have been detected. CBMs containing bound calcium include those from families 3, 4, 9 and 22. These metal ions appear to confer increased thermostability on the CBMs, consistent with their origin, which is generally from thermophilic microorganisms.

Inspection of the three dimensional structure of these proteins have started to reveal similarities between proteins from different families. Thus, CBM4 and CBM22 proteins have a very similar secondary structure with the high affinity calcium binding located in the same position in the two proteins. It is likely that, in future, CBMs will be subjected to a higher order of classification with similar families grouped into superfamilies or clans, analogous to the glycoside hydrolase clans described by Henrissat and Coutinho.

With the large amount of structural information now available on CBMs, the challenge of the next few years is to understand the structural basis for ligand specificity, particularly in those families where different members exhibit diverse ligand specificity. To some extent these studies have already started. For example, the orientation of the surface tryptophans in CBM2bs defines ligand specificity. Replacement of a surface arginine with a glycine re-orientated Trp 259 from a perpendicular to a planar orientation with respect to the other surface aromatic residue, Trp 291. The R262G mutant no longer bound to xylan but, instead, interacted with crystalline cellulose.[30] Similar studies on other CBM families are likely to reveal further insights into the structural basis for ligand specificity.

6   BIOLOGICAL IMPORTANCE OF CBMS

Clearly the presence of CBMs in an array of different degradative enzymes points to an important biological role for these proteins. Numerous studies have shown that cellulose binding CBMs potentiate cellulase activity against insoluble substrates, but has little effect on soluble substrates.[31] Similarly, cellulose binding CBMs, appended to xylanases, also enhance enzyme activity against insoluble cellulose/xylan composite structures such as unbleached kraft pulps.[32] It should be emphasized, however, that in the last few years it has become apparent that the ligands for CBMs are not restricted to various forms of cellulose but can encompass an array of different polysaccharides. It is not surprising, therefore, that there are numerous reports of CBMs enhancing the activity of glycoside hydrolases against polysaccharides such as insoluble starch, xylan and mannan.

The mechanism by which CBMs potentiate enzyme activity is a matter of some debate. It is highly likely that these proteins increase the rate of catalysis by bringing the enzymes into intimate and prolonged contact with its target substrate.[33] In addition,

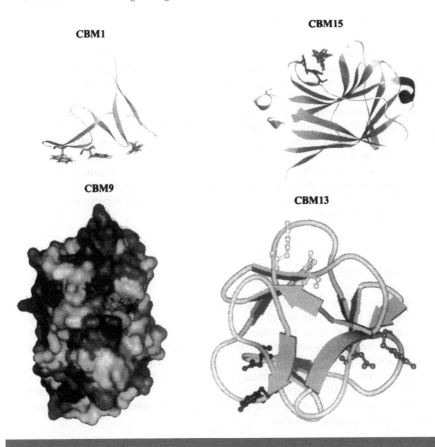

Figure 2 *The three dimensional structures of four new CBMs; a Type A (CBM10), a Type B (CBM15) and two Type C proteins (CBM9 and CBM13)*

electron microscopy and biochemical assays have indicated that the surface polysaccharide chains in cellulose microfibrils are disrupted by a *Cellulomonas fimi* CBM2a, leading to enhanced catalysis by increasing the accessibility of the substrate to enzyme attack.[34,35] This phenomenon of substrate disruption has not been reported for other CBMs, including *Pseudomonas cellulosa* CBM2as.[33] A different type of substrate disruption has, however, been described for CBMs that bind to starch. These modules contain two binding sites that can bind amylose chains. Biochemical studies, supported by Atomic Force microscopy has shown that these CBMs, by binding to two starch chains, disrupt the helical conformation of the polysaccharide, thus making it more accessible to enzyme attack.[36] An interesting question arising from the biological importance of CBMs is the evolutionary rationale for CBM9 proteins that bind to the ends of polysaccharide chains. The CBM may impact on the mode of action of the enzyme, such as introducing processivity from the reducing end of the polysaccharide or targeting to areas of the plant cell wall which have already been subject to enzyme attack.

7   BIOTECHNOLOGICAL APPLICATIONS

CBMs have considerable potential as affinity tags for the purification and/or immobilisation of fusion proteins. For immobilisation the 'irreversible' cellulose binders such as CBM2a modules are particularly appropriate. For protein purification the Type C CBMs have considerable potential as they can be eluted from insoluble polysaccharides by mono- or disaccharides. In addition, Type A CBMs can be modified so that elution from cellulose is possible. Replacing an aromatic residue on the ligand binding surface of a CBM1 generates a mutant protein whose affinity for cellulose mirrors the ionisation profile of the imidazole ring of histidine. Thus, a pH gradient can be used to elute the protein from the cellulose matrix.[37] One of the major advantages of using CBMs as affinity tags is that cellulose is the target ligand of many of these proteins. Cellulose is an extremely cheap, robust and inert matrix, and thus, in many ways represents an ideal affinity chromatography system.

8   CONCLUSIONS

Since the last Carbohydrate Bioengineering Meeting, considerable advances have been made in identifying CBMs with novel ligand specificities, and elucidating the three dimensional structures of these proteins. The challenge now is to understand the structural basis for the ligand specificity displayed by these modules. For example, some CBM families contain proteins that bind very different ligands, however, the structural basis for this diversity in carbohydrate recognition is unclear. Similarly, several CBM families that bind to crystalline cellulose contain a very similar ligand binding surface, which raises the question of how these proteins are capable of interacting with different regions of the crystalline polysaccharide.

References

1.   P. Tomme, R.A.J. Warren and N.R. Gilkes, *Adv. Microb. Physiol.*, 1995, **37**, 1.
2.   T. Watanabe, Y. Ito, T. Yamada, M. Hashimoto, S. Sekine, and H. Tanaka, *J. Bacteriol*, 1994, **176**, 4465.
3.   K. Sorimachi, M.F. Le Gal-Coeffet, G. Williamson, D.B. Archer and M.P. Williamson, *Structure,* 1997, **5**, 647.
4.   D. Stoll, A. Boraston, H. Stalbrand, B.W. McLean, D.G. Kilburn, and R.A.J. Warren, *FEMS Microbiol. Letts.,* 2000, **183**, 265.
5.   A.C. Fernandes, C.M.G.A. Fontes, H.J. Gilbert, G.P. Hazlewood, T.H. Fernandes and L.M.A. Ferreira, *Biochem. J.*, 1999, **342**, 105.
6.   P.M. Coutinho and B. Henrissat, 'Recent Advances in Carbohydrate Bioengineering', H.J. Gilbert, G.J. Davies, B. Henrissat and B. Svensson (Eds), The Royal Society of Chemistry, Cambridge, 1999, p. 3.
7.   A.B. Boraston, B.W. McLean, J.M. Kormos, M. Alam, N.R. Gilkes, C.A. Hayes, P. Tomme, D.G. Kilbrun and R.A.J. Warren, 'Recent Advances in Carbohydrate Bioengineering', H.J. Gilbert, G.J. Davies, B. Henrissat and B. Svensson (Eds), The Royal Society of Chemistry, Cambridge, 1999, p. 202.

8. P.M. Coutinho and B. Henrissat, http://afmb.cnrs-mrs.fr/~pedro/CAZY/cbm.html, 2000

9. G.Y. Xu, E. Ong, N.R. Gilkes, D.G. Kilburn, D.R. Muhandiram, M. Harris-Brandts, J.P. Carver, L.E. Kay, and T.S. Harvey, *Biochemistry,* 1995, **34,** 6993.

10. P.J. Simpson, D.N. Bolam, A. Cooper, A. Ciruela, G.P. Hazlewood, H.J. Gilbert and M.P. Williamson, *Structure,* 1999, **7,** 853.

11. A. Sunna, M.D. Gibbs, and P.L. Bergquist, *Biochem.J.,*2001, **356,** 791.

12. S.J. Charnock, D.N. Bolam, J.H. Turkenburg, H.J. Gilbert, L.M.A. Ferreira, G.J. Davies and C.M.G.A. Fontes. *Biochemistry,* 2000, **39,** 5013.

13. M. AbouHachem, E.N. Karlsson, E. BartonekRoxa, S. Raghothama, P.J. Simpson, H.J. Gilbert, M.P. Williamson, and O. Holst, *Biochem. J.,* 2000, **345,** 53.

14. J. Tormo, R. Lamed, A.J. Chirino, E. Morag, E.A. Bayer, Y., Shoham, and T.A. Steitz, *EMBO J.*, 1996, **15,** 5739.

15. B.W. McLean, M.R. Bray, A.B. Boraston, N.R. Gilkes, C.A. Haynes, D.G. Kilburn, *Prot. Eng.,* 2000, **13,** 801.

16. E.J. Jervis, C.A. Haynes and D.G. Kilburn, *J. Biol. Chem.,* 1997, **272,** 24016.

17. G. Carrard, A. Koivula, H. Soderlund, and P. Beguin, *Proc. Natl. Acad. Sci. U.S.A.,* 2000, **97,** 10342.

18. A.L. Creagh, E. Ong, E. Jervis, D.G. Kilburn and C.A. Haynes, *Proc. Natl. Acad. Sci. U.S.A.*, 1996, **93,** 12229.

19. P. Tomme, A.L. Creagh, D.G. Kilburn, and C.A. Haynes, *Biochemistry,*1996, **35,** 13885

20. H. Xie, D.N. Bolam, T. Nagy, L. Szabo, A. Cooper, P.J. Simpson, J.H. Lakey, M.P. Williamson and H.J. Gilbert, *Biochemistry*, 2001, **40,** 5700.

21. H. Xie, H.J. Gilbert, S.J. Charnock, G.J. Davies, M.P. Williamson, P.J. Simpson, S. Raghothama, C.M.G.A. Fontes, F.M.V. Dias, L.M.A. Ferreira and D.N. Bolam, *Biochemistry*, *in the press.*

22. M. Linder, I. Salovuori, L. Ruohonen and T.T. Teeri, *J. Biol. Chem,* 1996, **271,** 21268.

23. D.N. Bolam, H. Xie, P. White, P.J. Simpson, S.M. Hancock, M.P. Williamson, and H.J. Gilbert, *Biochemistry,* 2001, **40,** 2468.

24. S. Raghothama, P.J. Simpson, L. Szabó, T. Nagy, H.J. Gilbert and M.P. Williamson, *Biochemistry,* 2000, **39,** 978.

25. S.J. Charnock, D.N. Bolam, J.P. Turkenburg, H.J. Gilbert, L.M.A. Ferreira, G.J. Davies and C.M.G.A. Fontes, *Biochemistry,* 2000, **39,** 5013.

26. A.B. Boraston, A.L. Creagh, M.M. Alam, J.M. Kormos, P. Tomme, C.A. Haynes, R.A.J.Warren and D.G. Kilburn, *Biochemistry,* 2001, **40,** 6240.

27. V. Notenboom, A.B. Boraston, D.G. Kilburn and D.R. Rose, *Biochemistry,* 2001, **40,** 6248.

28. Z. Fujimoto, A. Kuno, S. Kaneko, S. Yoshida, H. Kobayashi, I. Kusakabe and H. Mizuno. *J. Mol. Biol.*, 2000, **300,** 575.

29. A.B. Boraston, P. Tomme, E.A. Amandoron, and D.G. Kilburn, *Biochem. J.*, 2000, **350,** 933.

30. P.J. Simpson, H. Xie, D.N. Bolam, H.J. Gilbert, H.J. and M.P. Williamson, *J. Biol. Chem.,* 2000, **275,** 41137.

31. P. Tomme, H. Van Tilbeurgh, G. Pettersson, J. Van Damme, J. Vandekerckhove, J. Knowles, T.T. Teeri and M. Claeyssens, *Eur. J. Biochem.*, 1988, **170,** 575.

32. G.W. Black, J.E. Rixon, J.H. Clarke, G.P. Hazlewood, M.K. Theodorou, P. Morris and H.J. Gilbert, *Biochem. J.*, 1996, **319,** 515.

33.   D.N. Bolam, A. Ciruela, S. McQueen-Mason, P.J. Simpson, M.P. Williamson, J.E. Rixon, A.B. Boraston, G.P. Hazlewood and H.J. Gilbert, *Biochem. J.*, 1998, **331**, 575.

34.   N. Din, H.G. Damude, N.R. Gilkes, R.C. Miller, R.A.J. Warren and D.G. Kilburn, *Proc. Natl. Acad. Sci. U.S.A.*, 1994, **91**, 11383-11387.

35.   N. Din, N.R. Gilkes, B. Tekant, R.C. Miller, R.A.J. Warren, and D.G. Kilburn, 1991, *Bio/Technol. 9*, 1096.

36.   S.M. Southall, P.J. Simpson, H.J. Gilbert, G. Williamson and M.P. Williamson MP., *FEBS Letts.*, 1999, **447**, 58.

37.   M. Linder, T. Nevanen and T.T. Teeri, *FEBS Letts.*, 1999, **447**, 13.

DOMAIN FUSION OF α-AMYLASE AND CYCLOMALTODEXTRIN GLUCANOTRANSFERASE

K. Ohdan and T. Kuriki

Biochemical Research Laboratory
Ezaki Glico Co., Ltd.
4-6-5 Utajima, Nishiyodogawa-ku
Osaka 555-8502, Japan

1 INTRODUCTION

α-Amylase (EC 3.2.1.1) and cyclomaltodextrin glucanotransferase (CGTase, EC 2.4.1.19), that are widely used in saccharifying industry, are interesting materials for basic research in improving the function of enzymes, because huge amount of information on these enzymes including their three-dimensional (3-D) structures have been stored up. Using various methods of protein engineering, several chimeric enzymes have been constructed out of amylolytic enzymes. Some of them have been studied with regard to secretion of the enzyme,[1] while others have been studied with regard to changes in substrate specificities,[2,3] product specificities,[4] or both.[5] However, it is difficult to donate a new function to a functional protein, which maintains its original function, because proteins are most likely composed of each part under strict rules, and some structural conflict occurs when new other proteins are introduced. It is generally known that hybrid proteins of different enzyme species do not express the functions derived from the original enzymes, since the original polypeptide-folding patterns are usually not maintained in the hybrid proteins. We report here one of the successful experiments of introducing raw starch-binding and -digesting ability of CGTase to an α-amylase. We also discuss at domain level on the relationship between structure and function of the chimeric enzymes.

2 CHOOSING BASE PROTEIN

We isolated a *Bacillus* strain that produced an unique type of α-amylase which strongly catalyzed transglycosylation reaction on hydroquinone and kojic acid.[6] These compounds have been reported to inhibit tyrosinase activity. Improving solubility of these compounds is very important from the viewpoint of industrial application. Actually, we are attempting to produce hydroquinone glucoside using transglycosylation reaction of this enzyme. The product, hydroquinone glucoside is most likely to be used for the inhibitor of tyrosinase. To improve the transglycosylation efficiency of the reaction system, we supposed that insoluble raw starch should be a better substrate than soluble starch, because there should be less acceptor in the reaction mixture except the objective acceptor, hydroquinone or kojic acid. It is also better to use raw starch as a substrate from the viewpoint of energy efficiency. We cloned the gene of this unique enzyme, and determined the nucleotide sequence and amino- (N-) and carboxyl- (C-) terminal amino acid sequences of the enzyme.[7] We found complete (Ba-L) and truncated (Ba-S) forms of the α-amylase and revealed that Ba-S was produced from Ba-L by proteolytical truncation of the 186 amino acid residues at the C-terminal region (Figure 1).

The characteristics of Ba-S were essentially the same as those of Ba-L,[7] which means that the extra 186 C-terminal amino acids of Ba-L were not necessary for the functionality of the α-amylase. Analysis of the secondary structure as well as the predicted 3-D structure of Ba-S demonstrated that Ba-S retained all of the domains (A, B, and C) which were most likely to be required for functionality as α-amylase.[7] From these data, we predicted that some functional domain could be easily introduced to the C-terminal region of Ba-S instead of the 186 amino acid residues polypeptide, without loosing α-amylase activity.

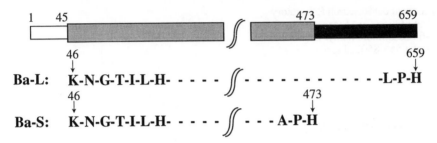

Figure 1 *N- and C-terminal amino acid sequences of Ba-L and Ba-S.*   White box indicates a signal peptide (45 amino acid residues) of the α-amylase.   Gray box corresponds to mature protein of Ba-S. Black box shows the C-terminal truncated polypeptide (186 amino acid residues) of Ba-L

3 CHOOSING DONOR DOMAINS

We cloned a gene of a new CGTase of alkalophilic *Bacillus* sp. A2-5a (A2-5a CGT) which is favorable for transglycosylation of many flavonoids[8] and analyzed the nucleotide sequence. We also found a typical sequence motif for raw starch-binding[9] at the C-terminal region, and showed the direct evidence of raw starch-binding ability.[10] It is generally known that a raw starch-binding domain (domain E) of CGTase should be an independent domain.[11-13] The domain maintains its original conformation and retains its starch-binding ability, even when the raw starch-binding domain is separated from the other four domains (A, B, C, and D).[13] Therefore, we adopted the raw starch-binding domain of A2-5a CGT. The amino acid sequence of A2-5a CGT was aligned with those of *Bacillus circulans* strain no. 8 CGTase[14] and *B. circulans* strain 251 CGTase[15] to determine the starting points of each domains in A2-5a CGT. The 3-D structures of the latter two enzymes have already been determined by X-ray crystallographic analysis.

4 DESIGN OF CHIMERIC ENZYMES

We designed chimeric enzymes made out of the α-amylase and the CGTase. Ba-S was used for the catalytic part of chimeric enzymes. We designed Ch1 Amy, which was composed of Ba-S and domain E of CGTase (Figure 2). The function of domain D of CGTase which connects domains C and E is unclear. It may be possible that domain D acts as a linker between catalytic domains (A, B, and C) and a raw starch-binding domain (E), as reported in glucoamylases.[16,17] Therefore, we also designed Ch2 Amy, which was composed of Ba-S and both D and E domains of A2-5a CGT (Figure 2). Chimeric genes were constructed by PCR fusion procedure.[18,19]

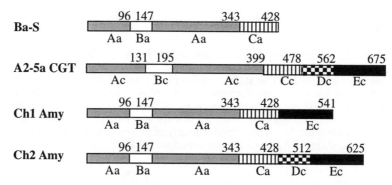

Figure 2 *Schematic diagram of Ba-S, A2-5a CGT, and chimeric enzymes.* Aa, Ba, and Ca indicate domains A, B, and C of Ba-S, respectively. Ac, Bc, Cc, Dc, and Ec indicate domains A, B, C, D, and E of A2-5a CGT, respectively. Numbers between domains show the last amino acid residues of each domain calculated from the N-terminus of mature enzymes

5   EXPRESSION OF HYBRID GENES AND CHARACTERIZATION OF CHIMERIC ENZYMES

The chimeric enzymes, Ch1 Amy and Ch2 Amy, were successfully produced in *Escherichia coli* .[20]   The molecular weights of Ch1 Amy and Ch2 Amy deduced from their nucleotide sequences were 59,514 and 67,701 respectively (Table 1), which were consistent with the molecular mass estimated by SDS-PAGE of the purified enzymes. Western analysis immunologically proved that the chimeric enzymes consisted of polypeptides from Ba-S and A2-5a CGT.[20]   We compared the characteristics of Ba-S and the chimeric enzymes.   There were no differences with regard to their pattern of action on soluble starch, optimum pHs, optimum temperatures, thermal stabilities, and transglycosylation reaction.   While Ba-L, Ba-S, and Ch1 Amy exhibited almost identical specific activities when enzyme activity was evaluated on a molar basis, the activity of Ch2 Amy was about one eighth of those of the other three enzymes (Table 1).

Table 1 *Specific activities and molar catalytic activities*

Enzyme	Specific activity (U/mg)	Molecular weight[a]	Molar catalytic activity[b] (U/nmol)
Ba-L	362	67,445	24.4
Ba-S	514	47,227	24.2
Ch1 Amy	420	59,514	25.0
Ch2 Amy	44.7	67,701	3.03

[a] Molecular weights were deduced from the nucleotide sequence

[b] Activities were evaluated on a molar basis

6   RAW STARCH-BINDING AND -DIGESTING ABILITIES OF CHIMERIC ENZYMES

The rate of adsorption to raw starch was measured (Table 2).   While Taka-amylase A (TAA, negative control) and Ba-S were hardly adsorbed to raw starch, Ch1 Amy and Ch2 Amy were adsorbed to raw starch.   Adsorption rate of Ch2 Amy was higher than that of Ch1 Amy and was almost the same as those of A2-5a CGT and porcine pancreas α-amylase (PPA, positive control). The ratios of raw starch-digestion by Ch1 Amy (39.5% at 36-h) and Ch2 Amy (56.6% at 36-h) were also much higher than that of Ba-S (11.3% at 36-h) (Figure 3).   These results clearly indicated that the raw starch-binding and -digesting abilities of A2-5a CGT were introduced to Ba-S.

Table 2 *Adsorption rate to raw starch (%)*

Ba-S	CGT[a]	Ch1 Amy	Ch2 Amy	PPA[b]	TAA[c]
9.4	68.5	49.8	63.0	76.7	3.2

[a]CGTase from alkalophilic *Bacillus* sp. A2-5a
[b]α-Amylase from porcine pancreas
[c]α-Amylase from *Aspergillus oryzae*

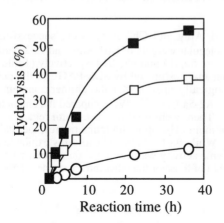

Figure 3 *Digestion of raw maize starch by Ba-S (O), Ch1 Amy (□), and Ch2 Amy (■).*   The same amounts of enzymes, Ba-S, Ch1 Amy, and Ch2 Amy, were used on a soluble starch-hydrolyzing activity basis.   The degree of hydrolysis (%) is indicated in terms of reducing sugars as glucose per total carbohydrate

7   INTERPRETATION OF RELATIONSHIP BETWEEN STRUCTURE AND FUNCTION IN CHIMERIC ENZYMES

Three important factors might have played a role in our success to produce enzymatically active hybrid proteins.   First, the folding of Ba-S protein is correct enough to retain its function as an α-amylase with or without the following extra C-terminal polypeptide.

Second, there are several amino acid residues that may link two functional proteins, as in glucoamylases.[16,17] Indeed, we identified a flexible structure at the C-terminus of the predicted 3-D structure of Ba-S.[7] Third, domain E of A2-5a CGT is functionally independent of other domains.

In spite of the remarkable decrease in the molar catalytic activity of Ch2 Amy (Table 1), it showed higher raw starch-binding and -digesting abilities than Ch1 Amy. One interpretation for these results could be as follows. Domain E of Ch1 Amy and domains D and E of Ch2 Amy could be folded in the same pattern as those of *B. circulans* strain 251 CGTase. We employed the 3-D structure of *B. circulans* strain 251 CGTase for A2-5a CGT, since their amino acid sequences are very similar (58.4 % homology). Since Ba-S and CGTases belong to the α-amylase family,[21-23] it is quite reasonable to think that the domains of chimeric enzymes maintain similar interactions between other domains as in CGTases, and that the domain(s) following domain C of chimeric enzymes are in a location similar to those of *B. circulans* strain 251 CGTase. When a 3-D structural model of Ba-S[7] was overlaid with the 3-D structure of *B. circulans* strain 251 CGTase, it was obvious that the loop between β4 and α4 of Ba-S collided with domain E at the main-chain level[20] (Figure 4).

Figure 4 *Simulation of introducing D plus E domains to Ba-S.*

This structural conflict causes some conformational change in the catalytic domain (Figure 4) and decreases the catalytic activity of Ch2 Amy (Table 1). On the other hand, it is highly probable that domain E of Ch1 Amy has little effect on the catalytic domain since domain E of Ch1 Amy is most likely to occupy the location of domain D of the CGTase. Indeed, a mobile segment with a *B*-factor above 40 Å^2 is located between domain C and domain D of the CGTase, indicating that the domain following domain C can behave in a flexible manner and settle into a suitable position with respect to the stability of the whole protein.[20]   This may explain why the molar catalytic activity of Ch1 Amy was similar to that of Ba-S (Table 1).

Ch2 Amy exhibited higher raw starch-binding and -digesting abilities than Ch1 Amy (Table 2, Figure 3).   Following interpretation can be possible; domain E of Ch2 Amy was in a better location, similar to that of the CGTase, than that of Ch1 Amy for binding to the substrate and expressing its function.   Thus, it is possible that domain D of CGTase may affect the location of domain E so that it is functionally optimized.   Another interpretation could be simply that a high concentration of domain E in the Ch2 Amy-reaction mixture, compared with that of Ba-S or Ch1 Amy, to adjust the amount of enzymes on starch-hydrolyzing activity basis disrupted water aggregates surrounding raw starch and that this made it easier for the catalytic domain to attack the hydrated starch micelle.[24]

It was reported that truncation of domains D and E of the CGTase from *Thermoanaerobacterium thermosulfurigenes* EM1 caused a remarkable decrease in its β-cyclomaltodextrin-forming activity and an increase in its saccharifying activity, indicating that domain D or domain E of CGTase is related to its transglycosylation activity.[11] Maltogenic α-amylase from *Bacillus stearothermophilus,* Novamyl, exhibits a five-domain organization extremely similar to CGTases.   The enzyme could be converted into a CGTase-like enzyme by the deletion of a small loop and substitution of two amino acid residues in domain B.[25]   Therefore, we conclude here that domains D and E of CGTase are not directly related to transglycosylation activity.   Indeed, we increased the transglycosylation activity of neopullulanase, one of the α-amylase family enzymes, by increasing the hydrophobicity along the entrance path of the attacking water molecule, which is most likely used for the hydrolysis reaction.[26]

References

1. N. Juge, M. Søgaard, J. C. Chaix, M. F. Martin-Euclaire, B. Svensson, G. Marchis-Mouren and X. J. Guo, *Gene,* 1993, **130**, 159.

2. H. Mori, K. Tanizawa and T. Fukui, *J. Biol. Chem.,* 1993, **268**, 5574.

3. M. Terashima, M. Hosono and S. Katoh, *Appl. Microbiol. Biotechnol.,* 1997, **47**, 364.

4. Y. J. Nakano and H. K. Kuramitsu, *J. Bacteriol.,* 1992, **174**, 5639.

5. T. Kuriki, D. C. Stewart and J. Preiss, *J. Biol. Chem.,* 1997, **272**, 28999.

6. T. Nishimura, T. Kometani, H. Takii, Y. Terada and S. Okada, *J. Ferment. Bioeng.,* 1994, **78**, 31.

7. K. Ohdan, T. Kuriki, H. Kaneko, J. Shimada, T. Takada, Z. Fujimoto, H. Mizuno and S. Okada, *Appl. Environ. Microbiol.,* 1999, **65**, 4652.

8. T. Kometani, Y. Terada, T. Nishimura, H. Takii and S. Okada, *Biosci. Biotechnol. Biochem.,* 1994, **58**, 1990.

9. B. Svensson, H. Jespersen, M. R. Sierks and E. A. MacGregor, *Biochem. J.,* 1989, **264**, 309.

10. K. Ohdan, T. Kuriki, H. Takata and H. Okada, *Appl. Microbiol. Biotechnol.,* 2000, **53**, 430.

11. R. D. Wind, R. M. Buitelaar and L. Dijkhuizen, *Eur. J. Biochem.,* 1998, **253**, 598.

12. A. Tanaka, S. Karita, Y. Kosuge, K. Senoo, H. Obata and N. Kitamoto, *Biosci. Biotechnol. Biochem.*, 1998, **62**, 2127.
13. B. Dalmia, K. Schütte and Z. Nikolv, *Biotechnol. Bioeng.*, 1995, **47**, 575.
14. C. Klein and G. E. Schulz, *J. Mol. Biol.*, 1991, **217**, 737.
15. C. L. Lawson, R. Vanmontfort, B. Strokopytov, H. J. Rozeboom, K. H. Kalk, G. E. Devries, D. Penninga, L. Dijkhuizen and B. W. Dijkstra, *J. Mol. Biol.*, 1994, **236**, 590.
16. T. Semimaru, M. Goto, K. Furukawa and S. Hayashida, *Appl. Environ. Microbiol.*, 1995, **61**, 2885.
17. G. Williamson, N. J. Belshaw and M. P. Williamson, *Biochem. J.*, 1992, **282**, 423.
18. J. Yon and M. Fried, *Nucleic Acids Res.*, 1989, **17**, 4895.
19. S. Fujiwara, H. Kakihara, K. B. Woo, A. Lejeune, M. Kanemoto, K. Sakaguchi and T. Imanaka, *Appl. Environ. Microbiol.*, 1992, **58**, 4016.
20. K. Ohdan, T. Kuriki, H. Takata, H. Kaneko and S. Okada, *Appl. Environ. Microbiol.*, 2000, **66**, 3058.
21. T. Kuriki and T. Imanaka, *J. Gen. Microbiol.*, 1989, **135**, 1521.
22. H. Takata, T. Kuriki, S. Okada, Y. Takesada, M. Iizuka, N. Minamiura and T. Imanaka, *J. Biol. Chem.*, 1992, **267**, 18447.
23. T. Kuriki and T. Imanaka, *J. Biosci. Bioeng.*, 1999, **87**, 557.
24. S. M. Southall, P. J. Simpson, H. J. Gilbert, G. Williamson and M. P. Williamson, *FEBS Lett.*, 1999, **447**, 58.
25. L. Beier, A. Svendsen, C. Andersen, T. P. Frandsen, T. V. Borchert and J. R. Cherry, *Protein Eng.*, 2000, **13**, 509.
26. T. Kuriki, H. Kaneko, M. Yanase, H. Takata, J. Shimada, S. Handa, T. Takada, H. Umeyama and S. Okada, *J. Biol. Chem.*, 1996, **271**, 17321.

STRUCTURE OF THE CATALYTIC MODULE AND FAMILY 13 CARBOHYDRATE BINDING MODULE OF A FAMILY 10 XYLANASE FROM *STREPTOMYCES OLIVACEOVIRIDIS* E-86 IN COMPLEX WITH XYLOSE AND GALACTOSE

Zui Fujimoto[1], Atsushi Kuno[2,3], Satoshi Kaneko[4], Hideyuki Kobayashi[4], Isao Kusakabe[2] and Hiroshi Mizuno[1,2]

[1]Department of Biochemistry, National Institute of Agrobiological Sciences, Tsukuba 305-8602 Japan
[2]Institute of Applied Biochemistry, University of Tsukuba, Tsukuba 305-8572 Japan
[3]Department of Material and Biological Chemistry, Yamagata University, Yamagata 990-8560, Japan
[4]National Food Research Institute, Tsukuba 305-8642, Japan

1 INTRODUCTION

Endo-1, 4-β-D-xylanase (xylanase, EC 3.2.1.8) hydrolyses β-1,4-glycosidic bonds within xylan, which is a major component of hemicelluloses in plant cell walls. Based on the amino acid sequence of the catalytic domains, xylanases have been classified mainly into two families, 10 and 11, of the glycoside hydrolases[1]. X-ray structure analyses show that the catalytic domains of family 10 xylanases consist of an eight-stranded $(\beta/\alpha)_8$-barrel with an active cleft running through the surface at the C-terminal side of the central β-barrel[2-4], while those of family 11 xylanases consist of β-sheet structures[5,6].

Besides the catalytic domain, xylanases frequently have a substrate-binding domain at the N- or C-terminal end to assist in the catalysis. Substrate-binding domains of xylanases are often specific for xylan, and are thus referred to as xylan-binding domains (XBD). Similarly, cellulose-binding domains (CBDs) are found in cellulases[7]. XBD, CBD, and other sugar-binding domains are now referred to as carbohydrate-binding modules (CBMs)[8]. CBMs are classified into more than 20 families based on amino acid sequence similarity. They constitute rather small domains (~120 amino acids), but have various types of structures[9].

Streptomyces olivaceoviridis E-86 xylanase (FXYN) has been used to produce xylobiose and xylose from commercial hardwood[10]. DNA sequence of FXYN indicates that it consists of a catalytic domain, a substrate-binding domain and a Gly/Pro-rich linker region connecting two domains, and the enzyme belongs to the glycoside hydrolase family 10[11]. The properties of FXYN have been studied using biochemical and structural approaches[11-14]. The XBD of FXYN has three series of repeated sequences and this type of XBD belongs to CBM family 13. Removal of the XBD from FXYN resulted in a truncated enzyme with about half the hydrolytic activity against insoluble xylan as the wild-type xylanase. This suggests that the XBD binds to insoluble xylan and assists in catalysis.

We have succeeded in the crystallization and structure determination of FXYN containing both the catalytic domain and XBD[15]. This was the first crystal structure of a xylanase containing an XBD. Family 13 CBM consists of three similar repeated peptides of about 40 residues each referred to as subdomains α, β, and γ, and these three units assemble around the pseudo three-fold axis, forming a globular structure. This combination of three subdomains results in a fold similar to the "β-trefoil fold" proposed by Murzin *et al*[16]. This fold is shared by plant galactose-binding lectins, ricin B-chain and abrin B-chain, in which two sets of domains with triple-repeated-sequence are arranged in tandem[17,18]. Comparison with the ricin/lactose complex showed that most of the residues involved in lactose binding in ricin are strictly conserved among all three subdomains of XBD, indicating that these sites are the major candidates for xylan binding.

Thus far, two studies have been conducted on the characteristics of the family 13 CBM[19,20]. To clarify the substrate-binding mode of XBD, we carried out structural analyses of the substrate complexes of FXYN using xylose (X1) and galactose (Gal), as well as xylobiose (X2) and lactose (Lac), as binding ligands.

2 METHODS

Recombinant full length FXYN was expressed in *Escherichia coli*, and purified by a previously described method[12]. Enzyme was crystallized by the hanging drop vapor diffusion method at room temperature using a 20 mg/ml protein solution and a reservoir solution composed of 25 or 30 % ammonium sulfate and 2% McIlvaine buffer (a mixture of 0.1M citric acid and 0.2 M Na_2HPO_4, pH 5.7)[21]. After a week, crystals grew to about 1 mm in length.

Crystals of X1, X2, galactose and lactose complexes of FXYN were prepared by carrying out the soaking experiments. Substrate was dissolved in the reservoir solution at a concentration of 20 mg /ml and crystals were soaked in the substrate solution. Diffraction measurements were made on a RIGAKU imaging-plate diffractometer RAXIS IV[++]. The data sets were processed and scaled using program CRYSTALCLEAR (RIGAKU). All the crystals belonged to the orthorhombic space group $P2_12_12_1$ with cell constants of about a=80, b=95, c=141. Structural analysis was initiated with the noncomplexed FXYN structure as the initial model. The resultant F_{obs}-F_{calc} and $2F_{obs}$-F_{calc} maps yielded the electron density corresponding to the soaked substrate. Sugar models were added into the model and successive manual model rebuilding was conducted. All structural models resulted in an R-factor of 17~20% and R_{free}-factor of 19~23% in the resolution range of 30-2.0 Å.

3 STRUCTURE OF THE SUGAR COMPLEX

All crystals included two non-crystallographic-symmetry (NCS)-related molecules (molecules A and B) and the root-mean-square (rms) differences for all atoms between two NCS molecules were below 1.0 Å. Therefore, descriptions hereafter are made on the molecule A. The overall structures and main-chain conformations were almost identical to those of the non-complexed enzyme. The rms differences between the complexes and non-complex structures were below 0.8 Å. The model of FXYN comprised of an N-terminal 301-residue $(\beta/\alpha)_8$-barrel structure, which was the catalytic domain and the 123-residue C-terminal ricin-type lectin domain that consisted of the XBD. The Gly/Pro-rich linker region between the two domains was not observed in the electron density maps and seems to be flexible in solution.

The electron densities corresponding to the soaked sugars were found in the catalytic cleft of the catalytic domain and in the sugar-binding site of subdomain α and γ of the XBD (Figure 1). Two NCS FXYN molecules were related by a non-crystallographic two-fold axis and were dimerized in the crystal. They interacted between the catalytic domain and XBD, and between the two XBDs, and the pseudo two-fold axis passed through between the NCS XBDs. As a result, the sugar-binding site of subdomain β of the XBD become positioned in the crystal packing interface between the NCS molecules and the soaked substrate seemed incapable of reaching the binding site.

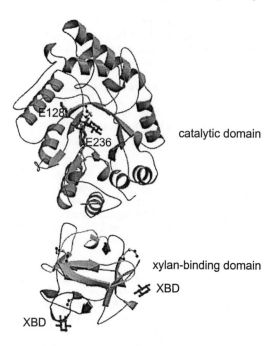

catalytic domain

xylan-binding domain

XBD

XBD

Figure 1 *Ribbon model of FXYN complexed with xylobiose. Two catalytic residues, disulfide bonds, and soaked xylobiose are shown. Linker region between 'c' and 'n' was not visible in the electron density map*

3.1 Crsytal structure of FXYN complexd with xylooligosaccharides

In the structures of the FXYN/X2 complex, two xylose sugar units were clearly interpreted from the electron density map to be at subsites –1 and –2, whereas in the complex of FXYN/X1, no clear density for glucose was observed in the catalytic cleft, even though high xylose concentrations were employed in the soaking experiments.

In the case of XBD, electron density corresponding to one xylose unit was observed in the subdomains α and γ in the complexes of FXYN/X1 and FXYN/X2 (Figure 2). These sites correspond to the galactose-binding sites of ricin B-chain[17]. Most residues involved in galactose binding in the ricin/lactose complex are well conserved and participate in xylose binding in FXYN. The O2 and O3 atoms of the xylose molecule were positioned in the inner part of the site, hydrogen-bonded to the side-chains of both Asp and Asn. O4 atoms made hydrogen bonds with Gln338 in subdomain α or Glu421 in subdomain γ and were at the XBD surface. Tyr340 and Tyr423 made stacking interactions with the xylose rings. There were also hydrogen bonding interactions between O3 atom of xylose and the main-chain carbonyl of Pro327 in subdomain α and Val410 in subdomain γ, and between O2 atom and the side-chain of His343 in subdomain α. In subdomain γ, a water molecule mediated between O2 atom of xylose and the side-chain oxygen of Ser426. The binding mode of xylose was different from that of galactose in ricin/lactose complex where the lactose ring is located perpendicular to the ricin B-chain with the O3 and O4 atoms buried in the site.

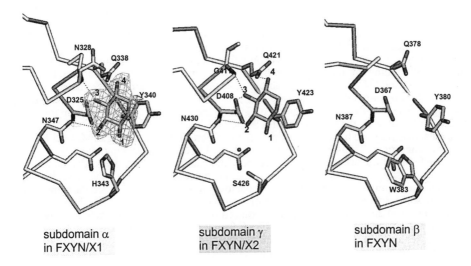

subdomain α
in FXYN/X1

subdomain γ
in FXYN/X2

subdomain β
in FXYN

Figure 2 *Xylose-binding structure at subdomain α in FXYN/X1 complex with the Fo-Fc omit electron density map (left), at subdomain γ in FXYN/X2 complex (center), and the structure of subdomain β in FXYN (right)*

The O1-O4 direction of xylose unit could not be determined from the 2Å resolution electron density maps because of the highly symmetrical structure of the xylose moiety. Besides the ring to ring stacking interaction with Tyr residues, C5 and O5 atoms had no polar interaction with the amino acid residues of the protein. Excluding these two atoms, xylose had a completely symmetrical structure with four equatorial hydroxyl groups. As it was difficult to discriminate between C5 and O5 atoms from the electron density maps, direction of the bound xylose was not strictly determined. These facts support the possibility that xylan binds in reverse direction at the binding sites. If the xylan-binding site could bind to the xylan chain in both directions, it would be advantageous for the enzyme. When XBD binds xylan, the catalytic domain could have a flexible position with respect to the XBD/xylan complex as the linker region is unstructured. Assuming that the role of XBD is only to bind substrate and not to direct the substrate into the active site of the enzyme, the above feature would increase the probability of the catalytic domain binding the substrate and thus effecting hydrolysis of the glycosidic bonds.

In the FXYN/X2 complexes, continuous electron densities were observed on both sides of the xylose unit bound through the O1 and O4 atoms in subdomain γ, and the XBD seems to bind any region of xylan if the substrate is not branched. As mentioned above, the xylan-binding site of subdomain β was buried in the crystal packing interaction and we could not observe the bound sugar. However, subdomain β had similarities in structure with subdomains α and γ, and for this reason we believe that it is likely to bind xylose (Figure 2). Thus, it seems that the XBD uses three xylose-binding sites to bind the xylan chain, thereby increasing the possibility of substrate binding.

3.2 Crystal structure of FXYN soaked with galactose and lactose

In the FXYN/lactose complex, an electron density corresponding to only one galactose unit of the lactose molecule was observed at the xylan-binding site of subdomain α (Figure 3). However, a continuous electron density from the O1 atom of the galactose moiety was found, which could correspond to the C4 atom of glucose, indicating that the bound sugar was lactose and the glucose unit might be flexibly positioned in the crystal. The electron density was also observed in subdomain γ, but it was too obscure to identify the bound molecule. In the FXYN/galactose, an electron density corresponding to the galactose unit was observed at the xylan-binding site of both subdomains α and γ. The manner of binding of the galactose moiety was similar to that in the ricin/lactose complex. O3 and O4 atoms of the galactose molecule were positioned in the inner part of the site, hydrogen-bonded to the side-chain of Asp325. The O3 atom had two more hydrogen bonds with the side-chains of His343 and Asn347. The O4 atom formed hydrogen bonds with the side-chain of Gln338 and the main-chain carbonyl of Asn328. The galactose ring had a chair conformation and a part of the sugar ring C3-C4-C5 faced the aromatic ring of Tyr343, where they participated in a partial stacking interaction.

In the FXYN/lactose complex structure, only the galactose moiety was observed in the electron density, whereas both galactose and glucose units are observed in the ricin/lactose complex. Orientation of the galactoses in FYYN and ricin are the same, but there are some differences in the manner of their binding. For example, in subdomain 1α of the ricin B-chain, Asn328 of XBD is replaced by Asp25, which interacts with the O3 atom of the glucose unit. But, the specificity for lactose in both proteins may be due to subtle differences caused by the surrounding residues, while XBD binds to galactose in the same manner as the galactose-binding lectin.

4   SUGAR-BINDING MODE OF FAMILY 13 CBM

Even though, CBM13 binds to xylose and galactose in a different manner, there are common features in the sugar coordination. In the FXYN/galactose complex structure, carboxylate of Asp325 hydrogen bonds with O3 and O4 atoms of the galactose moiety,

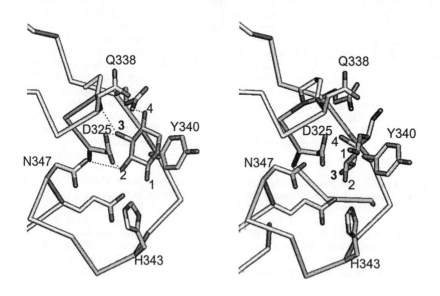

Figure 3 *Bound xylose and galactose in the subdomain α of XBD in the structures of FXYN/X1 and FXYN/Gal complexes.*

whereas it coordinates with O2 and O3 atoms of the xylose unit in FXYN/X1 or FXYN/X2 complex (Figure 3). In those complexes, the structure of O3-C3-C4-O4 of galactose and O2-C2-C3-O3 of xylose are placed at the same position. Since the O4 hydroxyl group of the galactose molecule protrudes in the axial position, the orientation of the sugar ring differs from that of the xylose molecule. Therefore, Asp325 in the subdomain α and Asp408 in the subdomain γ and their surrounding residues control the mode of binding and thereafter Tyr aromatic rings hold a part of the sugars.

When XBD binds to xylose, O1 and O4 atoms are positioned at the side of the binding pocket, such that the adjacent xylose units are linked by the β-1, 4-glycosyl bond so that it can bind to the long xylan chain. On the other hand, when XBD binds to galactose, the O1 atom of the galactose protrudes outwards. As a result, it can bind to the galactooligosaccharides in which galactose is linked to the other sugars through the O1 atom as in lactose, but not through the O4 atom. In this binding mode, the ricin B-chain binds to the terminal galactose or galactose derivative of the sugar chains at the cell surface, as dose the galactose-binding lectin[22].

Family 13 CBM binds at least two different sugars in a different manner and this mode of binding is effectively utilized by different enzymes.

References

1   B. Henrissat and G. Davies. *Cur. Opin. Struct. Biol.*, 1997, **7**, 637.
2   U. Derewenda, L. Swenson, R. Green, Y. Wei, R. Morosoli, F. Shareck, D. Kluepfel and Z. S. Derewenda. *J. Biol. Chem.*, 1994, **269**, 20811.
3   G. W. Harris, J. A. Jenkins, I. Connerton, N. Cummings, L. L. Leggio, M. Scott, G. P. Hazlewood, J. I. Laurie, H. J. Gilbert and R. W. Pickersgill. *Structure*, 1994, **2**, 1107.
4   A. White, S. G. Withers, N. R. Gilkes and D. R. Rose. *Biochemistry*, 1994, **33**, 12546.
5   A. Torronen, A. Harkki and J. Rouvinen. *EMBO J.*, 1994, **13**, 2493.
6   W. W. Wakarchuk, R. L. Campbell, W. L. Sung, J. Davoodi and M. Yaguchi. *Protein Sci.*, 1994, **3**, 467.
7   P. Tomme, R. A. J. Warren, R. C. Miller_Jr, D. G. Kilburn and N. R. GilkesC *Cellulose-binding domains: classification and properties* eds. J. M. Saddler andM. H. Penner American Chemical Society, Washington, DC, 1995.
8   P. Tomme, A. Boraston, B. McLean, J. Kormos, A. L. Creagh, K. Sturch, N. R. Gilkes, C. A. Haynes, R. A. J. Warren and D. G. Kilburn. *J. Chromatogr. B*, 1998, **715**, 283.
9   A. B. Boraston, B. W. McLean, J. M. Kormos, M. Alam, N. R. Gilkes, C. A. Haynes, P. Tomme, D. G. Kilburn and R. A. J. WarrenC *Carbohydrate-binding modules: diversity of structure and function* eds. H. J. Gilbert, G. J. Davies, B. Henrissat andB. Svensson The Royal Society of Chemistry, Cornwall, 1999.
10   S. Yoshida, T. Ono, N. Matsuo and I. Kusakabe. *Biosci. Biotech. Biochem.*, 1994, **58**, 2068.
11   A. Kuno, D. Shimizu, S. Kaneko, Y. Koyama, S. Yoshida, H. Kobayashi, K. Hayashi, K. Taira and I. Kusakabe. *J. Ferment. Bioeng.*, 1998, **86**, 434.
12   A. Kuno, D. Shimizu, S. Kaneko, T. Hasegawa, T. Gama, K. Hayashi, I. Kusakabe and K. Taira. *FEBS Letters*, 1999, **450**, 299.
13   S. Kaneko, A. Kuno, Z. Fujimoto, D. Shimizu, S. Machida, Y. Sato, K. Yura, M. Go, H. Mizuno, K. Taira, I. Kusakabe and K. Hayashi. *FEBS Letters*, 1999, **460**, 61.
14   S. Kaneko, S. Iwamatsu, A. Kuno, Z. Fujimoto, Y. Sato, K. Yura, M. Go, H. Mizuno, K. Taira, T. hasegawa, I. Kusakabe and K. Hayashi. *Protein Eng.*, 2000, **13**, 873.
15   Z. Fujimoto, A. Kuno, S. Kaneko, S. Yoshida, H. Kobayashi, I. Kusakabe and H. Mizuno. *J. Mol. Biol.*, 2000, **300**, 575.

16   A. G. Murzin, A. M. Lesk and C. Chothia. *J. Mol. Biol.*, 1992, **223**, 531.
17   E. Rutenber and J. D. Robertus. *Proteins Struct. Funct. Genet.*, 1991, **10**, 260.
18   T. H. Tahirov, T.-H. Lu, Y.-C. Liaw, Y.-L. Chen and J.-Y. Lin. *J. Mol. Biol.*, 1995, **250**, 354.
19   A. Kuno, S. Kaneko, H. Ohtsuki, S. Ito, Z. Fujimoto, H. Mizuno, T. Hasegawa, K. Taira, I. Kusakabe and K. Hayashi. *FEBS Letters*, 2000, **482**, 231.
20   A. B. Boraston, P. Tomme, E. A. Amandoron and D. G. Kilburn. *Biochem. J.*, 2000, **350**, 933.
21   Z. Fujimoto, H. Mizuno, A. Kuno, S. Yoshida, H. Kobayashi and I. Kusakabe. *J. Biochem.*, 1997, **121**, 826.
22   S. Olnes and A. Pihl. in *The Molecular action of toxins and viruses* eds. P. Cohen andS. V. Heynigen Elsevier, New York, 1982, p.52.

DESIGNER NANOSOMES: SELECTIVE ENGINEERING OF DOCKERIN-CONTAINING ENZYMES INTO CHIMERIC SCAFFOLDINS TO FORM DEFINED NANOREACTORS

Henri-Pierre Fierobe[1], Adva Mechaly[2], Chantal Tardif[1,3], Anne Belaich[1], Raphael Lamed[4], Yuval Shoham[5], Jean-Pierre Belaich[1,3] and Edward A. Bayer[2,*]

[1]Bioénergétique et Ingéniérie des Protéines, Centre National de la Recherche Scientifique, IBSM-IFR1, 13402 Marseille, France; [2]Department of Biological Chemistry, The Weizmann Institute of Science, Rehovot, Israel; [3]Université de Provence, Marseille, France; [4]Department of Molecular Microbiology and Biotechnology, Tel Aviv University, Ramat Aviv, Israel; [5]Department of Food Engineering and Biotechnology, and Institute of Catalysis Science and Technology, Technion—Israel Institute of Technology, Haifa, Israel

1 INTRODUCTION

The enzymatic systems of bacteria and fungi that hydrolyze complex carbohydrates, e.g., plant cell wall polysaccharides, comprise a collection of free enzymes, multifunctional enzymes, cellulosomes and/or related multi-enzyme complexes.[1-7] The enzymatic and other components of such systems are usually composed of modular proteins, in which the various functional modules or domains are attached together within a single polypeptide chain via intramolecular linking segments or linkers.[8-10] The modular nature of these components can be utilized as building blocks to construct chimeric cellulosome-like complexes,[11] hereby called "nanosomes". The nanosome concept is shown schematically in Figure 1.

1.1 Free Enzymes, Multifunctional Enzymes and Cellulosomes

In the relatively simple case of a free enzyme, the polypeptide chain would contain a catalytic domain that performs the actual enzymatic cleavage of the polysaccharide substrate and a separate carbohydrate-binding module or CBM that binds the enzyme to the substrate.[12,13] These two domains are usually connected by a linker. Commonly, the substrate of the enzyme is cellulose, the enzyme is a cellulase, and the CBM is a cellulose-binding domain, originally referred to as a CBD. Since cellulose is such a prominent and structurally important component of the plant cell wall, many enzymes, e.g., xylanases, that are not strictly cellulolytic in nature, may also contain a true CBD. In many cases, the polypeptide chain also includes other types of accessory domains that collectively serve to modulate the enzymatic activity of the catalytic domain. Again, the different modules of the complete enzyme are usually connected via a set of linking segments.

A multifunctional enzyme is a more intricate type of free enzyme that contains more than one catalytic domain in a single polypeptide chain.[14,15] The multiple catalytic domains often exhibit distinct types of enzymatic activity and belong to distinct glycosyl hydrolase families. In many cases, such enzymes also contain one or more CBMs and several other accessory domains, all of which jointly regulate the combined catalytic activities. As in the simple type of free enzyme, the series of modules that form the polypeptide chain in such an enzyme are commonly connected via linkers.

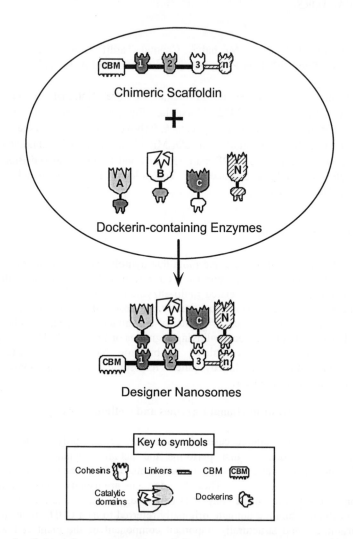

Figure 1 *Schematic representation of the nanosome concept.* A chimaeric scaffoldin is produced, containing one or more optional carbohydrate-binding modules (CBMs) and multiple (n) cohesin modules of different dockerin specificities. The dockerin counterparts comprise distinct modules as part of the polypeptide chains of the desired enzyme components. The nanosome complex is constructed by simply mixing in solution the chimeric scaffoldin and dockerin-containing components. The resultant designer nanosome exhibits enhanced synergistic functions, due to the close proximity of the interacting components.

Cellulosome-like complexes include more than one polypeptide chain.[16-23] In such complexes, one or more specialized "scaffoldin" subunits serve to organize the enzymes into a higher-order protein complex. For this purpose, the scaffoldins contain cohesin domains, usually in multiple copies, which interact selectively with dockerin domains of the enzyme subunits, thereby incorporating them into the complex. The intermolecular cohesin-dockerin interaction lays the basis for the construction of the cellulosome. In some cellulosomal systems, a scaffoldin bears its own dockerin domain, which binds to a different type of cohesin borne by an alternative scaffoldin-like, cell-surface molecule.

Initially, the genes of two scaffoldins from two different bacteria were sequenced. The first was termed "cellulose-binding protein A" (CbpA) from *Clostridium cellulovorans*.[24,25] The sequencing of such genes (proteins) was indeed a remarkable feat, owing not only to their very large size (e.g., ~1800 amino acid residues, ~5500 bp) but also to the presence of repeated sequences which complicates further positioning and reconstruction of the sequenced segments into the complete gene. At the time, it was not known what the internal repeating sequences represented (later shown to be 9 cohesins and 4 domains of unknown function) although a CBD was detected on the basis of sequence homology — hence the name of the protein.[25] A year later, the scaffoldin from *Clostridium thermocellum* was sequenced.[24] In this work, the nature of the 9 repeating units of this scaffoldin was realized, and the protein was termed "cellulosome-integrating protein A" or CipA. More recent work has provided complete sequences of several other scaffoldins from *Clostridium cellulolyticum, Clostridium josui, Acetivibrio cellulolyticus, Bacteroides cellulosolvens* and *Ruminococcus flavefaciens*, which range between 3 and 11 cohesin domains (Table 1).

1.2 Extension of the Cellulosome Concept

Several years ago, we proposed that the cellulosome system and its individual components can be utilized to construct modified types of cellulosome-like complexes.[11] Several approaches were suggested, including (a) formation of "supercellulosomes" via chemical or biochemical crosslinking of extraneous enzymes to intact, native cellulosomes to augment the intrinsic enzymatic activity of the complex; (b) formation of "heterocellulosomes" via incorporation of selected dockerin-containing enzymes into a native scaffoldin to regulate the content of enzymes within the cellulosome; and (c) design of chimeric scaffoldins and hybrid dockerin-containing enzymes to control precisely the composition and architecture of the resultant "chimeric cellulosome".

In the intervening years, the first two approaches have been explored to some extent with limited levels of success.[26] In many respects, however, the third strategy was ahead of its time. On the one hand, the concept of dismantling cellulosomal subunits to their individual functional elements afforded a conceptual basis for redesigning nature. By extrapolating, one could envisage the use of such a system for designing improved cellulosomes for more efficient management of cellulosic waste. On the other hand, our then-current knowledge of the structure, function and specificity of these building blocks was incomplete.

Table 1   *Currently sequenced scaffoldins*

Organism	Scaffol-din	Gene (bp)	Protein (aa res.)	CBD	No. of cohesins	Dockerin (C-term.)	Ref. No.
Acetivibrio cellulolyticus	CipV[a]	5748	1915	internal	7	Yes	27
Bacteroides cellulosolvens	CipBc	6951	2316	internal	11[b]	Yes	28
Clostridium cellulolyticus	CipC	4641	1546	*N*-term.	8	No	29
Clostridium cellulovorans	CbpA	5547	1848	*N*-term.	9	No	25
Clostridium josui	CipA (CipJ)	3489	1162	*N*-term.	6	No	30
Clostridium thermocellum	CipA	5562	1853	internal	9	Yes	24
Ruminococcus flavefaciens	ScaA	2640	879	none	3[c]	Yes	31
	ScaB	5259	1752	none	7[c]	No	31

[a]CipV also contains a unique *N*-terminal family-9 catalytic domain
[b]The cohesins of CipBc are of type-II.
[c]The cohesins of the *R. flavefaciens* scaffoldins have been classified as type-III

1.2.1 CBDs versus CBMs. At the time, it was generally accepted that CBDs were substrate-targeting modules,[8,32,33] and that cohesin and dockerin modules combined biochemically to incorporate the enzymes into the complex.[34,35] In many cases, sequence homology and group classification contributed to much of our then-current knowledge regarding these modules. Biochemical information regarding the activity of the different protein-carbohydrate or protein-protein interactions by individual modules or between modules was sparse. For example, several types of modules were sequenced, expressed and shown to exhibit cellulose-binding properties. These types of modules were then collectively termed CBDs. Later, when sequencing of cellulases and related proteins became more commonplace, similar modules were recognized on the basis of sequence homology and classified into a given CBD family, without necessarily examining the individual binding properties of the supposed CBD on cellulose or other polysaccharide substrates. Subsequently, it was determined that, despite the obvious sequence homology, the substrate specificity of some family members was indeed clearly for cellulose, whereas others preferred other substrates, e.g., xylan.[36] On the basis of these and similar findings, a decision was more recently made to extend the description of such modules, and the more general descriptive term carbohydrate-binding modules was adopted in lieu of cellulose-binding domains — i.e., CBM instead of CBD.[13] Nevertheless, the term CBD is still commonly retained when the preferred substrate is cellulose.

1.2.2 The Cohesin-Dockerin Interaction. Insight into the cohesin-dockerin interaction was first demonstrated for cellulosomal components in *Clostridium thermocellum* by Béguin and colleagues.[34,35,37,38] In their original work, it was shown by recombinant means that the enzyme dockerins recognize the scaffoldin cohesins, but the same type of cohesin failed to recognize its own scaffoldin-borne dockerin. Instead, the latter dockerin recognized a different type of cohesin, which characterized proteins that anchored the cellulosome to the cell-surface. The two types of cohesins and dockerins were similar but distinctive not only in their specificities but also in their sequences; those that incorporated the enzymes into the complex were termed type I and those involved in cell-surface anchoring were termed type II.[39]

When the original notions regarding chimeric cellulosomes were being considered, however, the fine points of the cohesin-dockerin interaction were still a mystery. For example, it was not known whether each cohesin within a given species of scaffoldin would recognize a specific dockerin (i.e., one cohesin — one dockerin) or whether some or all recognize the same set of dockerins. In the former case, we would expect that each cohesin would selectively incorporate a given enzyme into the complex. In the alternative case, the enzymes would be randomly incorporated into the complex (unless an alternative mechanism would account for selectivity of incorporation). Subsequent research served to clarify this enigma in *C. thermocellum*. By cloning and expressing individual recombinant cohesins from the scaffoldin, it was shown that the different cohesins all recognize the majority of the dockerin-containing cellulosomal enzymes in a similar manner.[40,41] It also became clear from several studies that the interaction was calcium dependent, a phenomenon that was traced to the presence of duplicated copies of a modified EF-hand motif.[40,42,43] Sequence homology indicated that the dockerins contained a calcium-binding loop in which all of the calcium-binding residues were clearly conserved. Calcium-binding experiments indicated that this loop was indeed responsible for the observed calcium dependence. Additional studies indicated that both duplicated segments of the dockerin were involved in cohesin binding.[44,45]

It was also unknown whether cohesins from one species would recognize dockerin-containing enzymes from another. Although each case (i.e., specificity characteristics among different species) must be analyzed individually, a pioneering study was first undertaken between cohesins and dockerins of *C. thermocellum* and *C. cellulolyticum*.[46] In this study, it was clearly demonstrated that the interaction between these two species was species specific — i.e., the cohesins of *C. thermocellum* failed to recognize the dockerins of *C. cellulolyticum* and *vice versa*. These findings together with comparative interspecies and intraspecies analyses of the various dockerin sequences allowed the proposal of a series of recognition determinants involved in the observed species specificity. Nevertheless, the precise molecular mechanism for species specificity has yet to be determined. In this context, the structures of recombinant cohesin domains from both *C. thermocellum*[47,48] and *C. cellulolyticum*[49] and a structural model for a *C. thermocellum* dockerin[50] have now been reported. However, we still lack a definitive structure of the cohesin-dockerin complex.

2 DESIGNER NANOSOMES: TAILOR-MADE CELLULOSOME CHIMERAS COMPOSED OF HYBRID COMPONENTS

The aim of the present work was to determine whether selective incorporation of enzymes into cellulosome-like complexes would result in increased synergistic action. For this purpose, a series of chimeric scaffoldins and hybrid enzymes were designed.

The chimeric scaffoldins comprised two cohesin domains of unlike specificity (one each from *C. thermocellum* or *C. cellulolyticum*) with or without one or two CBDs. Recombinant enzyme constructs contained a catalytic module together in the same polypeptide chain with a dockerin domain from either species. The nanosomes were assembled *in vitro,* simply by combining, in equimolar amounts, three desired components, i.e., the chimeric scaffoldin and two enzymes. The activity of the resultant nanosomes was examined using microcrystalline cellulose as a substrate and compared with that of the binary mixture of free enzymes.

2.1 Design and Preparation of Recombinant Nanosome Components

Five different chimeric scaffoldins were engineered (Table 2). Each construct contained two different cohesin species that exhibited divergent specificities[1]. Scaf1 and Scaf2 were based on the cellulosomal scaffoldin from *C. thermocellum,* in which the two cohesins are separated by an internal CBD[11]. Scaf3 is based on the *C. cellulolyticum* cellulosome and contains an N-terminal CBD[29]. Scaf4, which lacks a CBD, was designed to determine whether simple complexation of enzymes would also promote synergistic activity. Scaf5 contained two CBDs (one from each species) at the extremities of the molecule, together with two internal cohesins.

The enzyme components were all based on two well-established cellulosomal enzymes: the family-5 CelA[51] and the family-48 CelF[52] (herein referred to as Ac and Fc, respectively). To complement the native enzymes, two hybrid constructs (termed At and Ft) were designed, in which the intrinsic dockerin domain (designated "c") of the respective *C. cellulolyticum* enzyme was replaced by a dockerin domain (designated "t") of differing specificity. The latter dockerin was derived from the corresponding family-48 enzyme, CelS[53], from *C. thermocellum* (Table 2). Thus, four different enzyme pairs could be incorporated onto each chimeric scaffoldin: Ac + At, Ac + Ft, Fc + At, and Fc + Ft.

The engineered proteins were produced in *E. coli* and affinity-purified in one step on either cellulose (for CBD-containing constructs) or Ni-NTA (for His-tagged constructs).

2.2 Analysis of Nanosome Complexes

Complex formation in the presence of calcium was verified using three different techniques: native PAGE, gel-filtration HPLC and surface plasmon resonance (SPR). Native PAGE clearly demonstrated that near-complete complex formation could be achieved by simply mixing the nanosome components *in vitro.* Binary or ternary mixtures of the free proteins resulted in a single major band of altered electrophoretic mobility. These results were confirmed by gel filtration HPLC of stoichiometric mixtures of the same nanosome components. The peaks corresponding to the free components disappeared and were replaced by a major peak of higher molecular mass. Apparent masses of 150 kDa and 200 kDa were found for binary and ternary complexes, respectively. The results are in good agreement with the expected sizes of the desired nanosomes. Estimates of affinity parameters for the interaction of Ac and Fc with the various chimeric scaffoldins were performed by SPR using a Biacore apparatus. Similar K_d values of 1-2.5×10^{-10} M were measured. These results were in good agreement with previously determined values for the cohesin-dockerin interaction in *C. cellulolyticum*[44]. The affinity of the chimeric scaffoldins for At and Ft, however, was in both cases too

Table 2 *Nanosome components used in this study*

Nanosome component	Molecular mass	Modular notation (alphabetic)	Modular organization (symbolic)
Scaffoldins			
Scaf1	54,520 Da	**Coh*t*–CBD*t*–Coh*c***	
Scaf2	54,476 Da	**Coh*c*–CBD*t*–Coh*t***	
Scaf3	62,470 Da	**CBD*c*–x–Coh*c*–Coh*t***	
Scaf4	36,752 Da	**Coh*t*–Coh*c***	
Scaf5	84,422 Da	**CBD*c*–x–Coh*c*–Coh*t*–CBD*t***	
Enzymes			
Ac	51,527 Da	**CelA–Doc*c***	
Fc	78,443 Da	**CelF–Doc*c***	
At	51,660 Da	**CelA–Doc*t***	
Ft	80,559 Da	**CelF–Doc*t***	

Abbreviations used: **Coh*t*,** cohesin domain from *C. thermocellum*; **Coh*c*,** cohesin from *C. cellulolyticum*; **CelA** and **CelF,** catalytic domains from the respective *C. cellulolyticum* family-5 and family-48 enzyme; **Doc*t*** and **Doc*c*,** dockerin domain from *C. thermocellum* and *C. cellulolyticum*, respectively; x, domain of unknown function. Shaded symbols indicate cellulosomal components derived from *C. cellulolyticum* and unshaded symbols from *C. thermocellum*. ➞●, His tag.

high ($>10^{11}$ M^{-1}) to be determined using the Biacore system, suggesting a much stronger cohesin-dockerin interaction in the *C. thermocellum* cellulosome compared to that of *C. cellulolyticum*. The interaction between the chimeric scaffoldins and different combinations of the enzyme components indicated that complexation between the enzymes and the *C. cellulolyticum* cohesin does not disturb subsequent binding of the second enzyme to the *C. thermocellum* cohesin and *vice versa*. Moreover, the different species of cohesins were shown to bind in an independent manner to the appropriate dockerin-containing enzyme, irrespective of the order of incorporation.

2.3 Nanosomes Exhibit Enhanced Synergy

Nanosomes were generally found to be more active than the simple mixtures of the free enzyme pairs. Bringing two cellulolytic modules into close proximity clearly enhances the catalytic efficiency on crystalline cellulose. The observed enhancement of activity increased with incubation time (data not shown) and reached a maximum after 24 h. The cellulolytic activity of the enzyme complexes was further improved when the scaffoldin contained a CBD. When attached to the CBD-containing scaffoldins (Scaf1, Scaf2 and Scaf3), the heterogeneous enzyme mixtures (Ac + Ft or Fc + At) exhibited an enhanced synergy of about 3 to 4 fold. In the absence of a CBD (Scaf4), enhanced levels of synergy were also observed. The most effective combinations of components were Scaf1 and Scaf2 together with enzymes Fc and At. Interestingly, the homogeneous mixture of family-5 enzymes (Ac and At) also displayed levels of enhanced synergy, nearly equivalent to those of the heterogeneous mixture of Ac and Ft. In contrast, nanosomes containing only the family-48 enzymes (Fc and Ft) showed little if any synergy. It is also noteworthy that two CBDs in the chimeric scaffoldin `(Scaf5) served to inhibit the level of synergy.

Table 3   *Synergistic action of binary mixtures of enzymes organized into nanosomes via chimeric scaffoldins*

Chimeric Scaffoldin	*Binary Mixture of Enzymes*			
	Ac + At	**Fc + At**	**Ac + Ft**	**Fc + Ft**
Scaf1	2.5	3.6	3.0	1.4
Scaf2	2.7	4.0	3.0	1.45
Scaf3	2.6	3.1	2.4	1.2
Scaf4	1.3	1.9	1.6	1.1
Scaf5	1.8	2.4	1.9	1.3

Values represent the synergy factor of enzymatic activity as calculated by assessing the level of activity of the binary mixture of the designated enzymes in the presence of the given chimeric scaffoldin divided by that of the mixture of enzymes alone. Final concentrations of all enzymes and Scafs were set at 0.1 µM, and the concentration of microcrystalline cellulose was 3.5 mg ml^{-1}. The buffer was 20 mM Tris-maleate pH 6.0, 1 mM CaCl$_2$ (and 0.01% azide). Enzyme activity was determined after 24-h incubation of the desired preparation at 37°C, upon which the amount of soluble released sugars was measured. See Table 2 for definition of scaffoldins and enzymes.

3 CONCLUSIONS

In previous works, single enzyme components of the *C. thermocellum* cellulosome have been shown individually to exhibit enhanced activity on insoluble cellulose substrates upon incorporation via a suitable scaffoldin into a cellulosome-like complex.[54-57] In addition, a simplified cellulosome was reconstituted by combining purified preparations of native cellulosomal components, including the full-size scaffoldin with selected enzymatic subunits.[58,59]

In the present study, we investigated the enhanced synergistic interaction of a heterogeneous system, wherein two different recombinant cellulosomal enzymes were incorporated selectively into discrete artificial nanosome complexes by virtue of their vectorial interaction with defined chimeric scaffoldins. Each of the chimeric scaffoldins contained two cohesins of divergent specificities. The scaffoldins were designed to examine the contribution of location of the designated modules therein. Thus, Scaf1 and Scaf2 both contain an internal CBD and cohesins from the two species, but their position vis-à-vis the CBD is reversed. The content of Scaf3 is very similar to that of Scaf2, except its CBD is at an N-terminal rather than internal position. The cohesins of Scaf4 are identical to Scaf1, except Scaf4 lacks a CBD. Finally, Scaf5 was designed to contain two CBDs and was constructed using the relevant portions of Scaf1 and Scaf3.

The data indicated that the activity levels of all nanosomes were significantly higher than those of the combined free enzyme systems, thereby demonstrating that proximity of the different enzymes within the complex indeed appears critical to the observed enhancement of synergistic action. The presence of a targeting CBD in the chimeric scaffoldin conferred an additional contribution towards the final level of enzyme activity displayed by a given nanosome. However, two CBDs in the same scaffoldin appeared to inhibit the level of synergy, presumably due to the tighter binding of the complex to the substrate and consequent restrictions on lateral diffusion. Indeed, strong binding has been shown previously to inhibit the level of catalytic activity of the native cellulosome.

Some enzyme combinations proved better than others. It is currently unknown why nanosomes composed of the homogeneous mixture of Ac and At resulted in enhanced synergy while those of Fc and Ft displayed no synergy. It is also unclear why nanosomes containing the combination of Fc and At consistently showed heightened levels of synergy over those containing Ac and Ft. A possible stabilising effect of the cohesin-dockerin interaction on the Fc and/or At constructs could account for the observed differences.

In order to extend the system, improved levels of synergy may eventually be achieved by using higher-degree nanosome systems that contain additional or other combinations of enzymes. In this context, cohesin and dockerin pairs can be used from other cellulosome species[1] to selectively incorporate additional enzyme components into higher-order nanosomes. Our capacity to control the specific incorporation of enzymatic and non-enzymatic components into defined nanosome complexes should have considerable biotechnological value for a broad variety of applications.[11]

References
1   E. A. Bayer, H. Chanzy, R. Lamed and Y. Shoham, *Curr. Opin. Struct. Biol.,* 1998, **8**, 548.
2   P. Béguin and J.-P. Aubert, *FEMS Microbiol. Lett.,* 1994, **13**, 25.

3   P. Tomme, R. A. J. Warren and N. R. Gilkes, *Adv. Microb. Physiol.,* 1995, **37**, 1.
4   R. A. J. Warren, *Curr. Opinion Biotechnol.,* 1993, **4**, 469.
5   K. Ohmiya, K. Sakka, S. Karita and T. Kimura, *Biotechnol. Genet. Eng. Rev.,* 1997, **14**, 365.
6   H. J. Gilbert and G. P. Hazlewood, *J. Gen. Microbiol.,* 1993, **139**, 187.
7   D. B. Wilson and D. C. Irwin, *Adv. Biochem. Eng.,* 1999, **65**, 1.
8   N. R. Gilkes, B. Henrissat, D. G. Kilburn, R. C. J. Miller and R. A. J. Warren, *Microbiol. Rev.,* 1991, **55**, 303.
9   P. M. Coutinho and B. Henrissat, in *Genetics, biochemistry and ecology of cellulose degradation,* eds. K. Ohmiya, K. Hayashi, K. Sakka, Y. Kobayashi, S. Karita and T. Kimura, Uni Publishers Co., Tokyo, 1999, p. 15.
10   P. M. Coutinho and B. Henrissat, 1999.
11   E. A. Bayer, E. Morag and R. Lamed, *Trends Biotechnol.,* 1994, **12**, 378.
12   P. M. Coutinho and B. Henrissat, in *Recent advances in carbohydrate bioengineering,* eds. H. J. Gilbert, G. J. Davies, B. Henrissat and B. Svensson, The Royal Society of Chemistry, Cambridge, 1999, p. 3.
13   B. Henrissat, T. T. Teeri and R. A. J. Warren, *FEBS Lett.,* 1998, **425**, 352.
14   M. D. Gibbs, R. A. Reeves, G. K. Farrington, P. Anderson, D. P. Williams and P. L. Bergquist, *Curr Microbiol,* 2000, **40**, 333.
15   V. V. Zverlov, S. Mahr, K. Riedel and K. Bronnenmeier, *Microbiology,* 1998, **144**, 457.
16   E. A. Bayer, L. J. W. Shimon, R. Lamed and Y. Shoham, *J. Struct. Biol.,* 1998, **124**, 221.
17   E. A. Bayer, Y. Shoham and R. Lamed, in *Glycomicrobiology,* ed. R. J. Doyle, Kluwer Academic/Plenum Publishers, New York, 2000, p. 387.
18   R. Lamed and E. A. Bayer, *Adv. Appl. Microbiol.,* 1988, **33**, 1.
19   J.-P. Belaich, C. Tardif, A. Belaich and C. Gaudin, *J. Biotechnol.,* 1997, **57**, 3.
20   P. Béguin and M. Lemaire, *Crit. Rev. Biochem. Molec. Biol.,* 1996, **31**, 201.
21   R. H. Doi and Y. Tamura, *Chem. Rec.,* 2001, **1**, 24.
22   C. R. Felix and L. G. Ljungdahl, *Annu. Rev. Microbiol.,* 1993, **47**, 791.
23   S. Karita, K. Sakka and K. Ohmiya, in *Rumen microbes and digestive physiology in ruminants,* Vol. 14, eds. R. Onodera, H. Itabashi, K. Ushida, H. Yano and Y. Sasaki, Japan Sci. Soc. Press, Tokyo/S.Karger, Basel, 1997, p. 47.
24   U. T. Gerngross, M. P. M. Romaniec, T. Kobayashi, N. S. Huskisson and A. L. Demain, *Mol. Microbiol.,* 1993, **8**, 325.
25   O. Shoseyov, M. Takagi, M. A. Goldstein and R. H. Doi, *Proc. Natl. Acad. Sci. USA,* 1992, **89**, 3483.
26   E. A. Bayer and R. Lamed, *Biodegradation,* 1992, **3**, 171.
27   S.-Y. Ding, E. A. Bayer, D. Steiner, Y. Shoham and R. Lamed, *J. Bacteriol.,* 1999, **181**, 6720.
28   S.-Y. Ding, E. A. Bayer, D. Steiner, Y. Shoham and R. Lamed, *J. Bacteriol.,* 2000, **182**, 4915.
29   S. Pagès, A. Belaich, H.-P. Fierobe, C. Tardif, C. Gaudin and J.-P. Belaich, *J. Bacteriol.,* 1999, **181**, 1801.
30   M. Kakiuchi, A. Isui, K. Suzuki, T. Fujino, E. Fujino, T. Kimura, S. Karita, K. Sakka and K. Ohmiya, *J. Bacteriol.,* 1998, **180**, 4303.
31   S.-Y. Ding, M. T. Rincon, R. Lamed, J. C. Martin, S. I. McCrae, V. Aurilia, Y. Shoham, E. A. Bayer and H. J. Flint, *J. Bacteriol.,* 2001, **183**, 1945.

32   P. Tomme, R. A. J. Warren, R. C. Miller, D. G. Kilburn and N. R. Gilkes, in *Enzymatic degradation of insoluble polysaccharides*, eds. J. M. Saddler and M. H. Penner, American Chemical Society, Washington, D.C., 1995, p. 142.

33   D. M. Poole, E. Morag, R. Lamed, E. A. Bayer, G. P. Hazlewood and H. J. Gilbert, *FEMS Microbiol. Lett.*, 1992, **99**, 181.

34   K. Tokatlidis, S. Salamitou, P. Béguin, P. Dhurjati and J.-P. Aubert, *FEBS Lett.*, 1991, **291**, 185.

35   K. Tokatlidis, P. Dhurjati and P. Béguin, *Protein Eng.*, 1993, **6**, 947.

36   P. J. Simpson, H. Xie, D. N. Bolam, H. J. Gilbert and M. P. Williamson, *J. Biol. Chem.*, 2000, **52**, 41137.

37   S. Salamitou, K. Tokatlidis, P. Béguin and J.-P. Aubert, *FEBS Lett.*, 1992, **304**, 89.

38   S. Salamitou, O. Raynaud, M. Lemaire, M. Coughlan, P. Béguin and J.-P. Aubert, *J. Bacteriol.*, 1994, **176**, 2822.

39   E. Leibovitz and P. Béguin, *J. Bacteriol.*, 1996, **178**, 3077.

40   S. Yaron, E. Morag, E. A. Bayer, R. Lamed and Y. Shoham, *FEBS Lett.*, 1995, **360**, 121.

41   B. Lytle, C. Myers, K. Kruus and J. H. D. Wu, *J. Bacteriol.*, 1996, **178**, 1200.

42   S. Chauvaux, P. Béguin, J.-P. Aubert, K. M. Bhat, L. A. Gow, T. M. Wood and A. Bairoch, *Biochem. J.*, 1990, **265**, 261.

43   S. K. Choi and L. G. Ljungdahl, *Biochemistry*, 1996, **35**, 4906.

44   H.-P. Fierobe, S. Pagès, A. Belaich, S. Champ, D. Lexa and J.-P. Belaich, *Biochemistry*, 1999, **38**, 12822.

45   B. Lytle and J. H. D. Wu, *J. Bacteriol.*, 1998, **180**, 6581.

46   S. Pagès, A. Belaich, J.-P. Belaich, E. Morag, R. Lamed, Y. Shoham and E. A. Bayer, *Proteins*, 1997, **29**, 517.

47   L. J. W. Shimon, E. A. Bayer, E. Morag, R. Lamed, S. Yaron, Y. Shoham and F. Frolow, *Structure*, 1997, **5**, 381.

48   G. A. Tavares, P. Béguin and P. M. Alzari, *J. Mol. Biol.*, 1997, **273**, 701.

49   S. Spinelli, H. P. Fierobe, A. Belaich, J. P. Belaich, B. Henrissat and C. Cambillau, *J Mol Biol*, 2000, **304**, 189.

50   B. L. Lytle, B. F. Volkman, W. M. Westler, M. P. Heckman and J. H. Wu, *J Mol Biol*, 2001, **307**, 745.

51   H.-P. Fierobe, C. Gaudin, A. Belaich, M. Loutfi, F. Faure, C. Bagnara, D. Baty and J.-P. Belaich, *J. Bacteriol.*, 1991, **173**, 7956.

52   C. Reverbel-Leroy, S. Pagés, A. Belaich, J.-P. Belaich and C. Tardif, *J. Bacteriol.*, 1997, **179**, 46.

53   W. K. Wang, K. Kruus and J. H. D. Wu, *J. Bacteriol.*, 1993, **175**, 1293.

54   A. Ciruela, H. J. Gilbert, B. R. S. Ali and G. P. Hazlewood, *FEBS Lett.*, 1998, **422**, 221.

55   I. Kataeva, G. Guglielmi and P. Béguin, *Biochem. J.*, 1997, **326**, 617.

56   J. H. D. Wu, W. H. Orme-Johnson and A. L. Demain, *Biochemistry*, 1988, **27**, 1703.

57   M. Fukumura, A. Begum, K. Kruus and J. H. D. Wu, *J. Ferment. Bioeng.*, 1997, **83**, 146.

58   S. Bhat, P. W. Goodenough, M. K. Bhat and E. Owen, *Int. J. Biol. Macromol.*, 1994, **16**, 335.

59   M. K. Bhat, *Recent Res. Devel. Biotech. Bioeng.*, 1998, **1**, 59.

5 Chemo-enzymatic Carbohydrate Synthesis

CHEMI-ENZYMATIC SYNTHESIS OF TOXIN BINDING OLIGOSACCHARIDES

Y.R, Fang[1], K. Sujino[1], A. Lu[1], J. Gregson[2], R. Yeske[2], V.P. Kamath[2], R.M. Ratcliffe[2], M.J. Schur[3], W.W. Wakarchuk[3] and M.M. Palcic[1]

[1]Department of Chemistry, University of Alberta, Edmonton, AB Canada T6G 2G2

[2]SYNSORB Biotech, Inc., Suite 410, 1167 Kensington Crescent N.W., Calgary AB Canada T2N 1X7

[3]Institute for Biological Sciences, National Research Council of Canada, 100 Sussex Drive, Ottawa ON Canada K1A 0R6

1 INTRODUCTION

Microbes and microbial toxins have long been known to bind to glycoconjugates on host cell surfaces (1). Since binding is an early event in infection, a potential therapeutic approach is the inhibition of attachment using competitive oligosaccharide ligands. One limitation in this approach is that intrinsically weak protein-carbohydrate affinities must be overcome for therapeutic efficacy of low-molecular weight inhibitors (2). This can be overcome by utilizing multivalent glycoconjugates that collectively bind much more tightly than the corresponding monomeric species (2). This approach has been employed in two gastro-enteric applications where synthetic oligosaccharides are immobilized on silylaminated diatomite supports (3,4). Immobilized trisaccharide **1**, SYNSORB Pk® (Scheme 1) effectively binds shiga-like toxin produced by *Escherichia coli* O157:H7 the causative agent of hemorrhagic colitis and hemolytic-uremic syndrome (3). Immobilized SYNSORB Cd®, trisaccharide **2**, binds Toxin A produced by *Clostridium difficle*, the causative agent of pseudomembranous colitis and antibiotic-associated diarrhea (4). The supports effectively bind the respective bacterial toxins and neutralize them by preventing binding to intestinal cells.

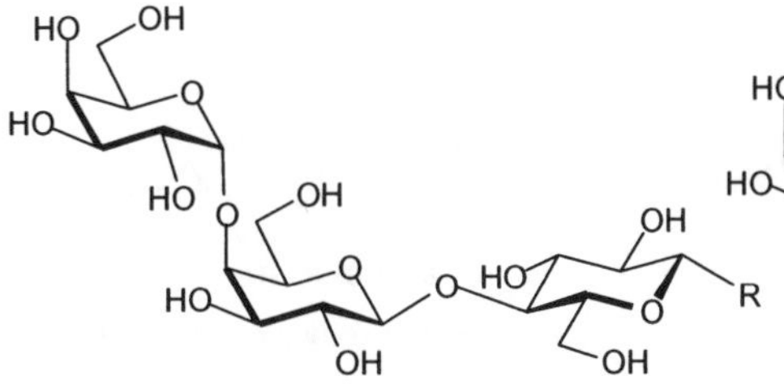

1 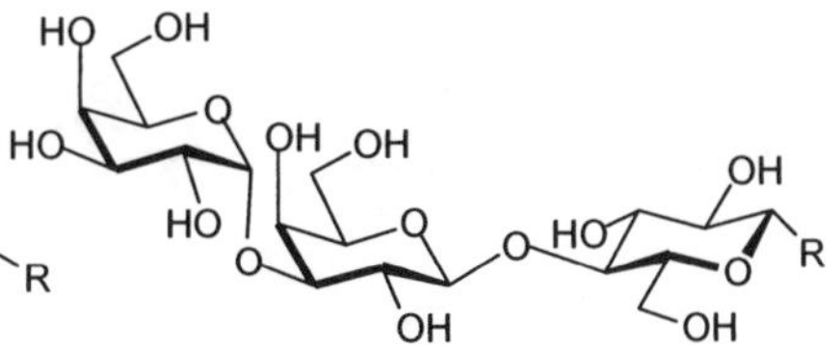   **2**

Scheme 1: The Structures of Toxin Binding Ligands. Compound **1**, SYNSORB Pk, binds shiga-like toxin of *E. coli* O157:H7; Compound **2,** SYNSORB Cd, binds Toxin A of *Clostridium difficle*.

The chemical synthesis of oligosaccharides is well established, however, it still remains a challenge despite the methodological advances that have been made over the last few decades (5-7). The hydroxyl or other functional groups on each saccharide building block must be selectively protected and then deprotected after coupling. Glycosylation generates a new stereogenic center at the anomeric carbon and no general methods are available to prepare all glycosidic linkages in a stereocontrolled fashion and in high yield. Chemical oligosaccharide synthesis is therefore a multi-step endeavor.

Biocatalysts are increasingly being employed in oligosaccharide synthesis. Several reviews on the use of enzymes in oligosaccharide synthesis have appeared recently (8-19). The two biocatalytic approaches use either biosynthetic glycosyltransferases or hydrolytic glycosidases. Glycosyltransferases catalyze the regio- and stereospecific transfer of a monosaccharide from a donor, frequently a nucleotide donor, to a variety of acceptors including saccharides. Multistep protection and deprotection sequences are thus avoided when using glycosyltransferases. Glycosidase reactions are also stereospecific and increasingly regiospecific. They can be used for formation of glycosides by reversal of their normal reactions under thermodynamic or kinetic control (8-19, 20).

The multi-kg scale chemical synthesis and immobilization of both trisaccharide ligands **1** and **2** have been carried out using lactose reducing sugar as a starting material. In this manuscript the chemi-enzymatic syntheses of Pk- and Cd trisaccharides from 8-methoxycarbonyloctyl-β-D-lactoside (**3**) on 50 g scale using glycosyltransferases are reported and compared to their preparation by chemical synthesis.

2 CHEMICAL SYNTHESIS

2.1 Synthesis of 8-Methoxycarbonyloctyl-β-D-Lactoside (3)

Lactose (**4**) was used as the starting material for the synthesis of 8-methoxycarbonyloctyl-β-D-lactoside (**3**), a key intermediate in both chemical and enzymatic trisaccharide synthesis. This aglycone is required for immobilization on silylaminated diatomite via their amides (21). Benzoylation of lactose (**4**) followed by reaction with HBr gave benzobromolactose **6**. Glycosylation of 8-carboxymethyloctanol with **6**, promoted by silver triflate then afforded the β-glycoside **7**. Zemplen de-O-benzoylation of **7** gave the desired lactoside **3** in > 80% yield at the multi-kg level.

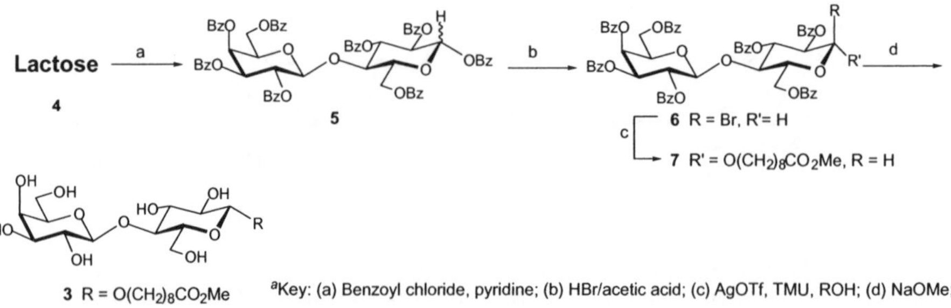

3 R = O(CH₂)₈CO₂Me   ᵃKey: (a) Benzoyl chloride, pyridine; (b) HBr/acetic acid; (c) AgOTf, TMU, ROH; (d) NaOMe;

Figure 1: The Synthesis of 8-Methoxycarbonyloctyl-β-D-Lactoside (3)

2.2 Chemical Synthesis of Pk Trisaccharide 1

The conversion of lactoside **3** to Pk trisaccharide **1** was achieved in 5 steps as shown in Figure 2. The 4',6'-O-benzylidene derivative of **3** was prepared and O-benzoylated to provide **8**. Regio-selective reductive ring-opening of the benzylidene ring in **8** then led to **9** where the required 4-OH group was available for glycosylation. Reaction of alcohol **9** with per-O-benzylated galactopyranosyl chloride, promoted again by silver triflate gave the expected α-linked galactoside **10**. The benzoyl esters and benzyl ethers were then conventionally deprotected to provide the required pK trisaccharide **1**. The overall yield of trisaccharide **1** from lactose (**4**) is 20-25% on kg scale.

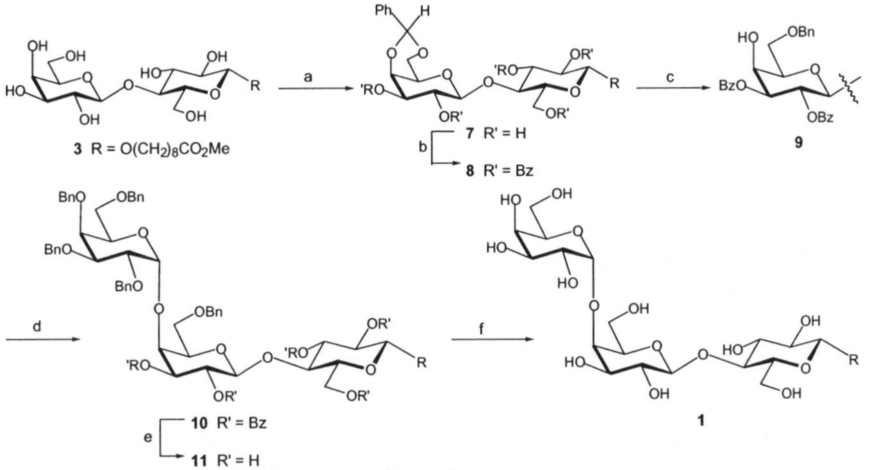

^{a}Key: (a) Benzaldehyde dimethyl acetal, p-TsOH; (b) Benzoyl chloride, pyridine (c) AlCl$_3$, BH$_3$.NEt$_3$, TFA, 0 ^{o}C; (d) 2,3,4,6 terta-O-benzyl-α-chloro galactoside, AgOTf, TMU; (e) NaOMe, MeOH: (f) H$_2$/Pd/C

Figure 2 . Chemical Synthesis of Pk Trisaccharide 1 from Lactoside 3.

2.3 Chemical Synthesis of *Clostridium difficile* Toxin A binding Trisaccharide 2

The common intermediate lactoside **3** was regioselectively converted to its 3'4'-O-isopropylidene derivative using dimethoxypropane as the acetalating agent to give **12** (Figure 3). Benzoylation followed by liberation of the 3',4'-OH groups by acidic hydrolysis then gave diol **14**. The orthoester of **14** was prepared by reaction with triethylorthoacetate and was regioselectively hydrolyzed to yield 4'-O-acetyl-3'-OH alcohol **15**. α-Galactosylation of **15**, using a per-O-benzylated thio-galactosyl donor, furnished the protected trisaccharide **16**, which was deprotected in 2 steps to yield the Cd-trisaccharide **2**. The overall yield of trisaccharide **2** from lactose (**4**) is 20-25% on kg scale.

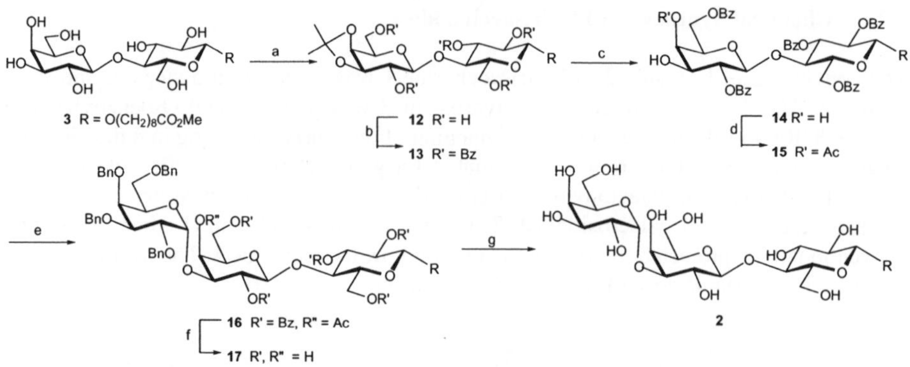

^aKey: (a) DMP, p-TsOH, 65 °C; (b) Benzoyl chloride, pyridine; (c) 80% acetic acid, 60 °C; (d) 1. triethylorthoacetate, p-TsOH; 2. 5% HCl; (e) 2,3,4,6 tetra-O-benzyl-β-thiobenzyl galactoside, CuBr₂, DMF, DCM; (f) NaOMe, MeOH; (g) H₂/Pd/C, MeOH

Figure 3. Chemical Synthesis of Cd Toxin A Binding Ligand (2) from Lactoside 3.

3   ENZYME SYNTHESIS

3.1   Chemi-Enzymatic Synthesis of Pk and Cd Toxin Binding Trisaccharides 1 and 2 using Glycosyltransferases and UDP-Gal Donor

Trisaccharides **1** and **2** can readily be synthesized from lactoside **3** using α1,4-galactosyltransferase (α1,4GalT) for Pk or α1,3-galactosyltransferase (α1,3GalT) for the CD toxin A binding ligand, respectively. These enzymes stereospecifically and regiospecifically transfer galactose from UDP-Gal donor to acceptor **3** (Figure 4). The α1,4GalT from *Neisseria meningitides* has been cloned and expressed in *E. coli* with an activity of ca. 500 Units/L of culture (22). Truncated bovine α1,3GalT from the FB3.Sp/Thy cell line was cloned into the plasmid pCWOri⁺ (23) with comparable levels of expression as those of α1,4GalT. One step purification on an SP-Sephadex fast-flow column yields functionally pure enzyme for use in synthesis.

Compound **1** was synthesized from 23 g of lactoside **3** using 350 Units of α1,4GalT and 1.2 equivalents of UDP-Gal donor in 1L. 3700 Units of alkaline phosphatase were also added to convert the UDP formed in the reaction to the less inhibitory uridine. Complete conversion to product took 38 hr at ambient temperature. Compound **2** was prepared from 47 g of lactoside 3 using 800 Units of α1,3GalT, 1.1 equivalents of UDP-Gal and 5000 Units of alkaline phosphatase in 2 L. This reaction was complete in 96 hr. After filtration to remove precipitate, products **1** and **2** were isolated in ca. 95% yield on reverse-phase C₁₈ columns.

3.2   Chemi-Enzymatic Synthesis of Cd Ligand 2 Using α1,3 Galactosyltransferase and UDP-Glc Donor

For cost reduction, UDP-Gal can be generated *in situ* from less costly UDP-Glucose with UDP-Gal epimerase (Figure 5).

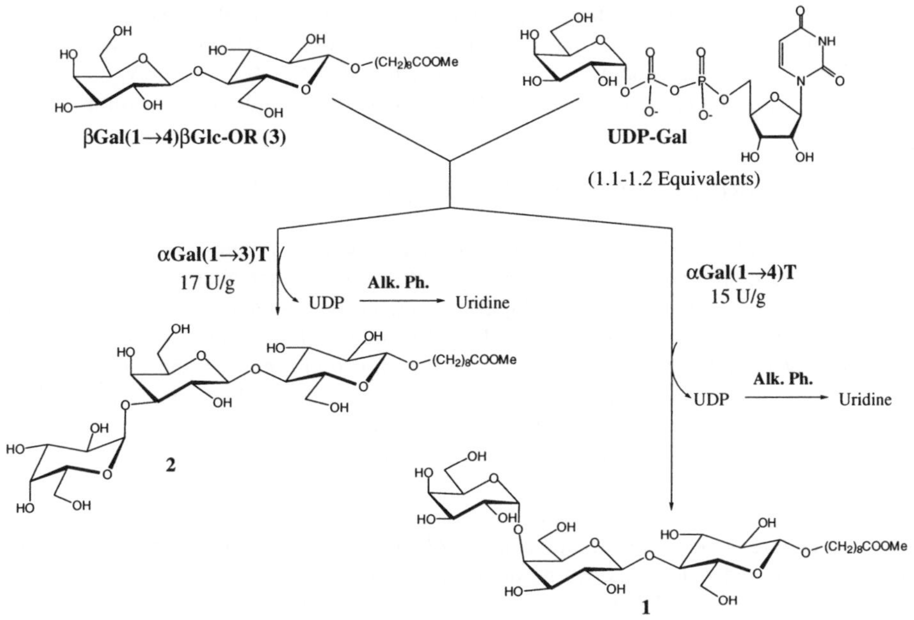

Figure 4 Enzymatic Synthesis of Toxin Binding Trisaccharides

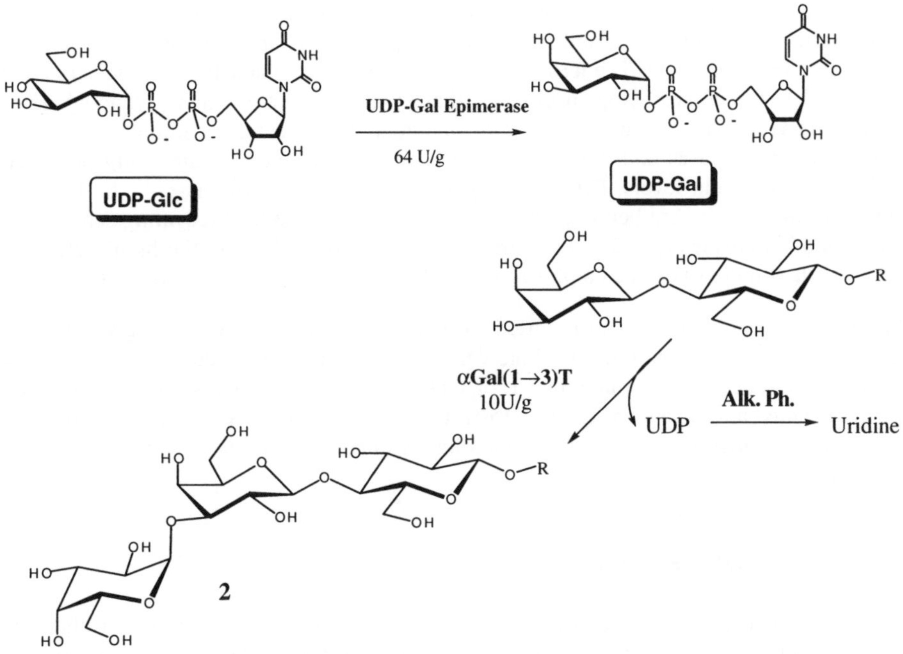

Figure 5 In situ Generation of UDP-Gal

Compound **2** was synthesized from 47g of lactoside **3** using 500 Units of α1,3GalT, 1.1 equivalents of UDP-Glc, 3000 Units of UDP-Gal epimerase and 6300 Units of alkaline phosphatase in 2 L. Reaction was complete in 112 hr. After filtration to remove precipitate, trisaccharide **2** was isolated in 95% yield on reverse-phase C_{18} columns.

4   COMPARISON OF CHEMICAL AND CHEMI-ENZYMATIC SYNTHESIS

The chemical synthesis of lactoside **3** with the appropriate linking arm for coupling to silylaminated diatomite gives a common precursor for further chemical or enzymatic elaboration. The synthesis of **3** is a highly optimized process starting with inexpensive lactose and is amenable to multi-kg GMP production with an overall yield >80%. Its further conversion to Cd and Pk trisaccharides as shown in Figures 1 and 2 give 20-25% yields from lactose on kg scale. Each kg of trisaccharide produces approx. 1500L of organic waste. The enzymatic routes will produce about 110 L organic waste/kg of trisaccharide. Yields on 50 g scale are >80% from lactose.

The enzymatic process requires high-level expression cloning of glycosyltransferases (500 U/L cell culture for lactoside acceptor) in *E. coli* or other scaleable hosts. UDP-Gal donor is costly, but cost reductions can be realized by using UDP-Glc as a donor along with UDP-Gal epimerase. UDP inhibition requires the addition of alkaline phosphatase to reactions. The enzymes are all stable at ambient temperature for at least one week.

5   FUTURE DIRECTIONS IN ENZYMATIC SYNTHESIS

The use of glycosyltransferases in oligosaccharide synthesis is increasing as limitations in their availability and the availability of the appropriate saccharide donors are overcome. Expression levels of 500 U/L microbial culture or more are being achieved. Bacterial and even viral enzymes are being discovered that synthesize mammalian glycoconjugates facilitating their cloning and overcoming mammalian biosynthetic limitations. Fusion proteins of transferases and donor synthetases or donor epimerases are being produced with high expression levels. Efficient recycling schemes that eliminate the build-up of inhibitory nucleotides are being investigated. Besides the classical recycling schemes of Wong and Whitesides (24, 25), the pyrophosphatase cycle (26) and the highly efficient Elling cycle (Figure 6) can be employed for UDP-Gal generation from sucrose (27).

Alternative methods have been reported for Cd and Pk synthesis *via* transglycosylation (28,29) and bacterial coupling of whole cells (30). The latter produced 188 g/L of Pk trisaccharide reducing sugar and 44 g/L of UDP-Gal in fermentation (31). Reaction rate enhancements and broadening of substrate specificity of glycosyltransferases by mutagenesis is ongoing. Chemi-enzymatic oligosaccharide synthesis on industrial scale is becoming feasible.

6   ACKNOWLEDGEMENTS

Funding from the Natural Sciences and Engineering Research Council of Canada is gratefully acknowledged. We thank Prof. P. G. Wang for the UDP-Gal epimerase clone.

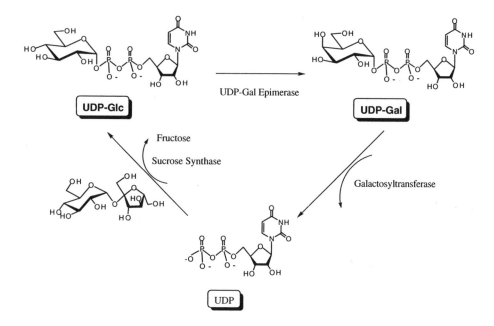

Figure 6. Efficient Recycling of Donor by the Method of Elling.

7 REFERENCES

1. K.-A. Karlsson, *Mol. Microbiol.*, 1998, **29**,1.
2. M. Mammen, S.-K. Choi and G.M. Whitesides, *Angew. Chem. Int. Ed.*, 1998, **37**, 2754.
3. G.D. Armstrong, P.C. Rowe, P. Goodyer, E. Orrbine, T.P. Klassen, G. Wells and P.N. McLain, *J. Infect. Dis.*, 1995, **171**, 1042.
4. L.D. Heerze, M.A. Kelm, J.A. Talbot and G.D. Armstrong, *J. Infect. Dis.,* 1994, **169**, 1291.
5. B. Ernst, G. W. Hart and P. Sinaÿ, *Carbohydrates in Chemistry and Biology* Vol. 1 Wiley-VCH Verlag, Weinheim, 2000.
6. G.J. Boons, Ed. *Carbohydrate Chemistry*, Blackie, London, 1998.
7. R.R. Schmidt and K.-H. Jung, *Carbohydr. Eur.*, 1999, **27**, 13.
8. S. Takayama, S. and C.-H. Wong, *Curr. Org. Chem.*, 1997, **1**, 109.
9. G.M. Watt, P.A.S. Lowden and S.L., Flitsch, *Curr. Opin. Struct. Biol.*, 1997, **7**, 652.
10. L. Elling, *Adv. Biochem Eng. Biotechnol.,* 1997, **58**, 89.
11. Z. Guo and P.G. Wang , *Appl Biochem Biotechnol* 1997, **6**,1.
12. U. Gambert and J. Thiem, *Top. Curr. Chem.,* 1997, **186**, 21.
13. D.H.G. Crout and G. Vic, *Curr. Opin. Chem. Biol.*, 1998, **2**, 98.
14. R. Ohrlein, *Top. Curr. Chem.,* 1999, **200**, 227.
15. J.M. Elhalabi and K.G. Rice, *Curr. Med. Chem.*, 1999, **6**, 93.
16. M.M. Palcic, *Curr. Opin. Biotech.*, 1999,**10**, 616.
17. N. Wymer and E.J. Toone, *Curr. Opin. Chem. Biol.,* 2000, **4**, 110.
18. K.M. Koeller and C.-H. Wong, *Chem. Rev.* 2000, **100**, 4465.

19.   B. Ernst, G.W. Hart and P. Sinaÿ, *Carbohydrates in Chemistry and Biology* Vol. 2, Wiley-VCH Verlag, Weinheim, 2000.

20.   M. Scigelova, S. Singh and D.H.G. Crout, *J. Mol. Catal. B, Enzym.,* 1999, **6**, 483.

21.   R.U. Lemieux, D.R. Bundle and D. A. Baker, *J. Am. Chem. Soc.*, 1975, **97**, 4076.

22.   W.W. Wakarchuk, A. Cunningham, D.C. Watson and N.M. Young, *Protein Eng.* 1998, **11**, 295.

23.   W.W. Wakarchuk, R.L. Campbell, W.L. Sung, J. Davoodi and M. Yaguchi, *Protein Sci.*, 1994, **3**, 467.

24.   C.-H. Wong, S.L. Haynie and G.M. Whitesides, *J. Org. Chem.*, 1982, **47**, 5416.

25.   C.-H. Wong, R. Wang and Y. Ichikawa, *J. Org. Chem.,* 1992, **57**, 4343.

26.   T. Noguchi and T. Shiba, *Biosci. Biotechnol. Biochem.,* 1998, **62**, 1594.

27.   T. Butler and L. Elling, *Glycoconjugate* J., 1999, **16**, 147.

28.   M. Scigelova and D.H.G. Crout, *J. Mol. Catal.. B, Enzym.,* 2000, **8**, 175.

29.   M. Scigelova and D.H.G. Crout, *Chem. Commun,* 1999, 2065.

30.   T. Endo and S. Koizumi, *Curr. Opin. Struct. Biol.,* 2000, **10**, 536.

31.   S. Koizumi, T. Endo, K. Tabata and A. Ozaki, *Nature Biotech.*, 1998, **16**, 847.

ENGINEERING THERMOSTABLE FAMILY 1 β-GLYCOSIDASES FOR SACCHARIDE PROCESSING

Thijs Kaper, John van der Oost, and Willem M. de Vos

Laboratory of Microbiology
Hesselink van Suchtelenweg 4
NL-6703 CT
Wageningen University
Wageningen
The Netherlands

1 INTRODUCTION

There is considerable interest in saccharide processing, which is primarily caused by the scientific challenges that are related to the understanding and use of the appropriate enzymatic systems. Additionally, there is an increasing interest of the food industry, where bulk compounds such as the milk sugar lactose are valorized into mono-, oligo- and polysaccharides that have potential as sweetners, biothickners, or neutraceuticals.[1] These food applications require food-grade production systems and, hence, significant attention has been given to lactic acid bacteria that couple the capacity to produce enzymes as well as polysaccharides to a long history of safe use. While the food-grade production and engineering of polysaccharides has been reviewed in considerable detail,[2-4] the use of hydrolases to process disaccharides, such as lactose, has received little attention. In this communication we discuss the use and engineering of thermostable family 1 β-glycosidases that have potential for use in industrial saccharide processing, can be produced easily in food-grade systems including lactic acid bacteria,[5] and provide novel insights in the structure-function of this group of hydrolases.

2 THERMOSTABLE FAMILY 1 GLYCOSYL HYDROLASES

Family 1 β-glycosidases can be used for hydrolysis of disaccharides and synthesis of oligosaccharides and glycoconjugates (Figure 1). β-Glycosidases from hyperthermophilic organisms, which have their optimum for growth above 80°C, offer several advantages over enzymes from mesophiles, such as functionality at elevated temperatures and increased operational stability. A high reactor temperature results in increased substrate solubility, increased reaction rates, and reduced risks of bacterial contamination.[6] The most thermostable family 1 β-glycosidases have been found in hyperthermophiles belonging to the Archaea. *Pyrococcus furiosus*, a marine organism with an optimum growth temperature at 100 °C,[7] and *Sulfolobus solfataricus*, isolated from acidic solfataric pools and optimally growing at 85 °C[8], both produce intracellular β-glycosidases that are capable of hydrolyzing a variety of β-linked saccharides with the

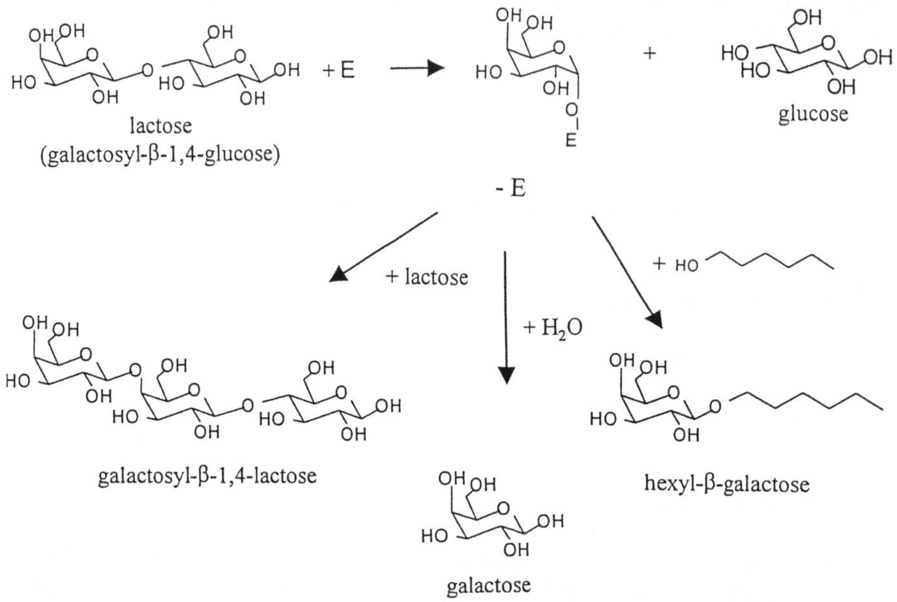

Figure 1 *Reactions catalyzed by family 1 β-glycosidases*

highest specificity for β-1,3-linked glucosides.[9-11] Both glycosidases belong to the glycosyl hydrolase family 1,[12,13] in which all domains of life are represented.[14]

The cognate family 1 glycosidases of hyperthermophilic Archaea can be divided in three separate groups (Figure 2), each of which contains one or more well-characterized enzymes. The *P. furiosus* CelB and *S. solfataricus* LacS belong to group A and have been subject of numerous structure-function studies.[15-19] Both enzymes are very thermostable and highly active, and have *in vivo* roles in the degradation of β-linked sugars, notably laminarin in *P. furiosus*[20] and most likely xyloglucan in *S. solfataricus* [21]. *P. furiosus* BmnA and *P. horikoshii* BglB of group B are β-mannosidases with low and very low hydrolytic activity, respectively.[5,13] The physiological role of these enzymes is unknown, but the specificity and low activity could suggest a role in biosynthesis or turnover of mannose containing compatible solutes.[22] The intracellular *P. horikoshii* BglA of group C is associated to the cytosolic membrane and has a high specificity for alkalic glucosides.[23]

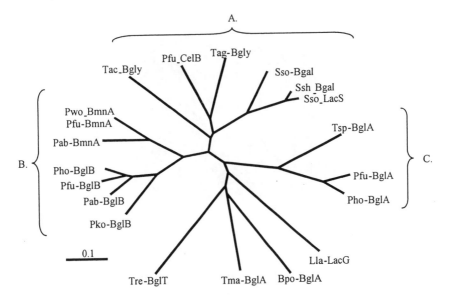

Figure 2 *Phylogenetic tree of Archaeal family 1 β-glycosyl hydrolase protein sequences, including Eukaryal and Eubacterial reference sequences. A. (putative) β-glycosidases: Pfu_CelB: P. furiosus β-glucosidase CelB (Genbank AF013169), Tag-Bgly: T. aggregans β-glycosidase (AF053078), Sso-LacS: S. solfataricus β-glycosidase LacS (M34696), Sso-Bgal: S. solfataricus β-galactosidase Bgal (X15950), Ssh-Bgal: S. shibatae β-galactosidase Bgal (L47841). B. (putative) β-mannosidases: Pho-BglB: P. horikoshii β-mannosidase (AP000002), Pab-BglB: P. abyssi putative β-mannosidase BglB (AJ248288), Pfu-BglB: P. furiosus putative β-mannosidase BglB (P. furiosus genome ORF Pf_368506), Tko-BglB: Thermococcus kodakaraensis putative β-mannosidase BglB (Genbank AB028601), Pfu_BmnA: P. furiosus β-mannosidase BmnA (U60214), Pab-BmnA: P. abyssi putative β-mannosidase BmnA (AJ248285). C: (putative) membrane associated β-glucosidases: Pfu-BglA: P. furiosus putative β-glucosidase BglA (AF195244), Pho-BglA: P. horikoshii β-glucosidase BglA (C71144), Tsp-BglA: Thermococcus sp. putative β-glucosidase BglA (Z70242), D. Bacterial and eukaryal β-glycosidases: Tre-BglT: Trifolium repens cyanogenic β-glucosidase (X56733), Tma-BglA: Thermotoga maritima β-glucosidase BglA (X74163), Bpo-BglA: Bacillus polymyxa β-glucosidase BglA (M60210), Lla-LacG: Lactococcus lactis 6-phospho-β-galactosidase LacG (M28357)*

The hyperthermophilic origin of the β-glycosidases is reflected by their extreme thermostability. This stability is intrinsic to the amino acid sequence and independent from the producing organism, since functional expression of the thermostable β-glycosidases in bacterial hosts, such as *Escherichia coli*[12,24] or *Lactococcus lactis*,[5], and eukaryal hosts like *Saccharomyces cerevisae*[25] yields enzymes with wild-type properties. The mesophilic hosts offer simple down-stream processing for purification: virtually pure protein can be obtained following heat incubation and clearing of cell-free extract by centrifugation.[26]

Figure 3 *Ribbon representation of tetrameric β-glucosidase CelB from* P. furiosus [15]

3 REACTION MECHANISMS AND 3D STRUCTURES

Family 1 glycosyl hydrolases hydrolyze their substrates with overall retention of the stereochemistry at the C1 atom. The experimentally deduced double-displacement mechanism has been confirmed by the presence and position of two essential, fully conserved glutamate residues in family 1 protein sequences and determined 3D structures.[14,27] The mechanism of catalysis close to 100 °C in hyperthermophilic family 1 enzymes is equal to that in their mesophilic counterparts at ambient temperatures.[12,28] The 9 amino acid residues, which have been shown to interact with the substrate in the – 1 subsite in crystal structures co-crystallized with ligands, have been almost fully conserved in sequence and position in the archaeal family 1 protein sequences.[15,17,29] An exception can be made for the group B BglB sequences (Figure 2), which have a unique active site structure that determines their catalytic profile.[30,31] The central fold of family 1 glycosyl hydrolases is the $(\beta\alpha)_8$-barrel,[32] which in the case of the determined structures of *S. solfataricus* LacS[17], *T. aggregans* Bgly[29] and *P. furiosus* CelB[15] has been arranged in a homotetramer (Figure 3). Organisation in oligomers has been observed in other thermostable enzymes as well and is believed to be a thermostablizing mechanism.[33,34]

The extreme thermostability of proteins from hyperthermophiles has attracted considerable attention, since it appears that thermostabilization is not a consequence of a general mechanism, but rather the result of a high number of locally optimized interactions.[34,35] In general, an increase in hydrogen bonds and ionic interactions, as well as an optimized packing of the hydrophobic core are observed in hyperthermostable proteins, compared to more thermosensitive counterparts.[33-35] Besides oligomerization to tetramers, the *P. furiosus* CelB and *S. solfataricus* LacS contain a relatively high number of ion-pairs at subunit interfaces.[17] Such charge interactions are regarded as an important mechanism for thermostabilization.[36]

4 SUBSTRATE SPECIFICITY AND ACTIVITY

The range of commercially interesting sugars for food or therapeutic applications consists of low-cost saccharides, such as fructose, glucose and lactose. Like the *S. solfataricus* LacS, the *P. furiosus* β-glucosidase CelB has considerable β-galactosidase activity, although CelB is most specific for the conversion of glucose disaccharides, such as cellobiose and laminaribiose.[9,37,38] CelB and LacS have potential for the hydrolysis of lactose at temperatures $\geq$ 70 °C.[39] This is still well below the optimal temperature for catalysis of both enzymes [9,10] and, consequently, CelB and LacS displayed long half-life times of activity. Under these conditions, however, CelB was found to be sensitive for substrate-induced inactivation at 70 °C.[39]

For optimal functioning in a non-natural setting, it might be desirable to change the properties of naturally occurring enzymes. Following structural comparison with related β-glycosidases, the substrate specificity of the β-glucosidase CelB has been probed by site-directed mutagenesis. The 6-phospho-β-galactosidase activity of CelB was increased after introduction of a phosphate-binding site in the catalytic center,[15] similar to the one present in the wild-type 6-phospho-β-galactosidase LacG from *Lactococcus lactis*.[40] The three introduced residues increased the catalysis rate (E417S), pH-optimum for hydrolysis (M424K) and specificity for galactosides (F426Y).[15] Likewise, CelB was modified for the hydrolysis of mannosides by comparison with the low-active β-mannosidase BglB from *P. horikoshii*. Two active site residues, located close to the catalytic residues, were exchanged between the enzymes. In CelB, the specificity for mannosides was increased by introduction of R77Q and N206D in combination an impaired stabilization of the transition state of the reaction. BglB was turned into a highly active β-glucosidase by introduction of D206N.[31]

Due to their complexity, rational design of enzymes is rarely successful.[41] In a directed evolution approach using error-prone PCR in combination with DNA shuffling,[42,43] β-glucosidase CelB has been optimized for conversion of para-nitrophenol-β-glucose at room temperature.[16] CelB mutant N415S displayed three times increased activities on the screening substrate at room temperature in combination with unaffected thermostability, showing that thermoactivity and thermostability are not necessarily coupled in thermostable enzymes. Alternatively, the recently developed DNA family shuffling technique[44,45] was employed to generate β-glycosidases more fit to hydrolyze lactose, without the limitations of suboptimal performance at a process temperature of 70°C and product inhibition.[39] The *P. furiosus* CelB and *S. solfataricus* LacS that share 56% nucleotide identity, were shuffled, and functional, thermostable hybrids were screened for increased hydrolysis of lactose at 70°C.[46] Selected high performance hybrids 11, 18, and 20 resulted from a single cross over near the N-terminus, which yielded proteins with N-terminal LacS sequence and C-terminal CelB sequence. The stability of the hybrids ($t_{\frac{1}{2},92 \,°C}$: 7-100 min) was intermediate of that of the parental enzymes ($t_{\frac{1}{2},92 \,°C}$: LacS: <3 min, CelB: >> 100 min), while they displayed several fold increased lactose hydrolysis rates compared to CelB (Figure 4). Advantage of enzyme optimization by directed evolution is the fact that structural information about the enzyme system of interest is not required. A combination of rational design and laboratory evolution for altering enzyme properties might be efficient and successful.[41]

5 SYNTHESIS OF OLIGOSACCHARIDES

Glycosidases can be employed for the synthesis of novel oligo-saccharides, which occurs as a side reaction during disaccharide hydrolysis through the retaining double-displacement mechanism.[47] A promising source for oligosaccharides is provided by the milk disaccharide lactose (galactosyl-β-1,4-glucopyranose).[48] The limited solubility of lactose at room temperature has initiated synthesis studies at elevated temperatures with the β-glycosidases CelB and LacS as catalysts. Both enzymes produce GOS in relatively high yields during hydrolysis of lactose.[30,39,49] The elevated temperatures resulted in higher lactose concentrations and subsequent higher GOS yields up to 43% w/w.[30] The optimum temperature for the GOS synthesis was well below the optimum temperatures of both enzymes, since above 75 °C the enzymes were rapidly inactivated due to Maillard reactions.[49] Both CelB and LacS have a preference for the formation of galactose-β-1,3/6-glucose bonds.[50] Besides the use of elevated temperatures, the high yields for CelB in GOS synthesis[30,49,50] is determined by the structure of its catalytic centers, since its yields could be improved by active site substitutions.[30]

By removal of the amino acid residue, which acts as the nucleophile in the synthesis and hydrolysis reactions, hydrolysis of the synthesis products is prevented. In this case however, the activity of the enzyme has to be restored by use of α-glycosylfluorides[51,52] or external nucleophiles[53,54]. When the catalytic nucleophile Glu387 was replaced by Ala or Gly, *S. solfataricus* LacS was completely inactivated. Activity was restored by addition of azide or formate, which in the case of formate yielded a saccharide synthetizing enzyme.[54] The LacS "glycosynthase" can be used for the synthesis of linear and branched oligosaccharides with β-1,3/4/6 linkages.[55]

6 GLYCOCONJUGATE SYNTHESIS

The coupling of sugars to aglycons yields products that find various applications in the

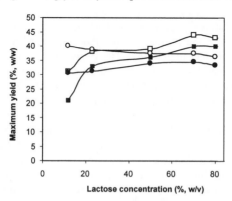

Figure 5. *Maximal oligosaccharide yield as a function of lactose concentration by the wild-type (■), the M424K mutant (●), the F426Y mutant (□), and the M424K/F426Y double mutant (O) of* P. furiosus *β-glucosidase CelB. Reaction conditions: 30U/ml glucosidase, 95 °C, pH 5.0.*[30]

soap, food, cosmetics or farmaceutical industry.[56] Through family 1 β-glycosidases the non-reducing moiety of a disaccharide can be coupled to alkylic chains (Figure 1).[56] Alkylic galactosides can be produced by *S. solfataricus* β-glycosidase and have potential as surfactants or components for sugar base polymers.[57] In contrast to mesophilic counterparts, the β-glucosidase CelB from *P. furiosus* is able to glucosylate tertiary alcohols by transglycoslylation.[58] Moreover, the CelB enzyme has been found to glucosylate hexanol via direct glucosylation.[59]

7 CONCLUSIONS

The archaeal thermostable family 1 β-glycosidases provide a promising and versatile tool for saccharide manipulation. With superior stability, high activity, and simple production, they can be easily employed on a large scale.[26] The vast amount of research done on substrate binding and hydrolysis mechanism of family 1 β-glycosidases is directly applicable to the thermostable family members and provides an excellent starting point for activity engineering studies.[15] Besides food-related applications like hydrolysis of lactose and oligo-saccharide synthesis, the extreme thermostable β-glucosidases can be used as biosensors or reporter enzymes.[60] The extreme thermostability of the proteins is an attractive feature for biotechnologists.[6] Similar as reported before,[61] the rigid (βα)$_8$-barrels might serve very well as scaffolds for the laboratory evolution of enzymes with new functions.

References

1. De Vos, W. M. et al., 1998, *Int. Dairy Journal* 8, 227-233.
2. Van Kranenburg, R., Boels, I. C., Kleerebezem, M. & De Vos, W. M., 1999, *Curr Opin Biotechnol* 10, 498-504.
3. Kleerebezem, M. et al., 1999, *Ant van Leeuwenh* 76, 357-65.
4. Boels, I. C., Van Kranenburg, R., Hugenholtz, J., Kleerebezem, M. & De Vos, W. M., 2001, *Int. Dairy J, in press.*
5. Kaper, T. et al., 2001 *Method Enzymol* 330, 329-46.
6. Niehaus, F., Bertoldo, C., Kähler, M. & Antraninkian, G., 1999, *Appl Microbiol Biotechnol* 51, 711-729.
7. Fiala, G. & Stetter, K. O., 1986, *Arch Microbiol* 145, 56-61.
8. Grogan, D. W., 1989, *J. Bacteriol.* 171, 6710-9.
9. Kengen, S. W., Luesink, E. J., Stams, A. J. & Zehnder, A. J., 1993, *Eur J Biochem* 213, 305-12.
10. Pisani, F. M. et al., 1990, *Eur J Biochem* 187, 321-8.
11. Kengen, S. & Stams, A., 1994, *Biocatalysis* 11, 79-88.
12. Voorhorst, W. G., Eggen, R. I., Luesink, E. J. & de Vos, W. M., 1995, *J Bacteriol* 177, 7105-11.
13. Bauer, M. W., Bylina, E. J., Swanson, R. V. & Kelly, R. M., 1996, *J Biol Chem* 271, 23749-55.
14. Henrissat, B., 2001, http://afmb.cnrs-mrs.fr/~pedro/CAZY/ghf.html.
15. Kaper, T. et al., 2000, *Biochemistry* 39, 4963-70.
16. Lebbink, J. H. G., Kaper, T., Bron, P., Van der Oost, J. & Vos, W. M., 2000, *Biochemistry* 39, 3656-65.
17. Aguilar, C. F. et al., 1997, *J Mol Biol* 271, 789-802.
18. Moracci, M., Capalbo, L., Ciaramella, M. & Rossi, M., 1996, *Protein Eng* 9, 1191-5.
19. D'Auria, S. et al., 1998, *Biochimie* 80, 949-57.
20. Voorhorst, W. et al., 1999, *J. Bact.* 181, 3777-3783.
21. Moracci, M. et al., 2000, *J Biol Chem* 275, 22082-9.
22. Da Costa, M. S., Santos, H. & Galinski, E. A. in *Advances in Biochemical Engineering Biotechnology* ed. Atranikian, G., 117-54 Springer-Verlag, Berlin,, 1998,.

23. Matsui, I. et al., 2000, *FEBS Lett* 467, 195-200.
24. Moracci, M. et al., 1995, *Enzyme Microb Technol* 17, 992-7.
25. Moracci, M., La Volpe, A., Pulitzer, J. F., Rossi, M. & Ciaramella, M., 1992, *J Bacteriol* 174, 873-82.
26. Lebbink, J. H. G., Kaper, T., Kengen, S. W. M., Van der Oost, J. & De Vos, W. M., 2001, *Meth Enzymol* 330, 364-79.
27. Kempton, J. B. & Withers, S. G., 1992, *Biochemistry* 31, 9961-9.
28. Bauer, M. W. & Kelly, R. M., 1998, *Biochemistry* 37, 17170-8.
29. Chi, Y. I. et al., 1999, *FEBS Letters* 445, 375-383.
30. Hansson, T., Kaper, T., Van der Oost, J., De Vos, W. M. & Adlercreutz, P., 2001, *Biotechnol Bioeng* 73, 203-10.
31. Kaper, T. PhD Thesis Wageningen University, Wageningen, The Netherlands, 2001.
32. Banner, D. W. et al., 1975, *Nature* 255, 609-14.
33. Vieille, C. & Zeikus, G. J., 2001, *Microbiol Mol Biol Rev* 65, 1-43.
34. Scandurra, R., Consalvi, V., Chiaraluce, R., Politi, L. & Engel, P. C., 1998, *Biochimie* 80, 933-41.
35. Jaenicke, R. & Bohm, G., 1998, *Current opinion in structural biology* 8 6,, 738-748.
36. Vetriani, C. et al., 1998, *Proc Natl Acad Sci USA* 95, 12300-5.
37. Driskill, L. E., Bauer, M. W. & Kelly, R. M., 1999, *Biotechnol Bioeng* 66, 51-60.
38. Nucci, R., Moracci, M., Vaccaro, C., Vespa, N. & Rossi, M., 1993, *Biotechnol Appl Biochem* 17, 239-50.
39. Petzelbauer, I., Nidetzky, B., Haltrich, D. & Kulbe, K. D., 1999, *Biotechnol Bioeng* 64, 322-32.
40. Wiesmann, C., Hengstenberg, W. & Schulz, G. E., 1997, *J Mol Biol* 269, 851-60.
41. Cedrone, F., Ménez, A. & Quéméneur, E., 2000, *Curr Opin Struct Biol* 10, 405-410.
42. Stemmer, W. P. C. *Nature*, 1994,370, 389-91.
43. Volkov, A. A. & Arnold, F. H., 2000, *Methods Enzymol* 328, 447-56.
44. Crameri, A., Raillard, S.-A., Bermudez, E. & Stemmer, W. P. C., 1998,*Nature* 391, 288-291.
45. Chang, C. C. J. et al., 1999, *Nat Biotechnol* 17, 793-7.
46. Kaper, T., Brouns, S. J. J., Geerling, A. C. M., De Vos, W. M. & Van der Oost, J., 2001, *submitted for publication*.
47. Watt, G. M., Lowden, P. A. S. & Flitsch, S. L., 1997, *Curr Opin Struct Biol* 7, 652-60.
48. Yang, S. T. & Silva, E. M., 1995, *J Dairy Sci* 78, 2541-62.
49. Boon, M. A., Van der Oost, J., De Vos, W. M., Jansen, A. E. M. & Van 't Riet, K., 1998,*Appl Biochem Biotechnol* 75, 269-78.
50. Petzelbauer, I., Zeleny, R., Reiter, A., Kulbe, K. D. & Nidetzky, B., 2000, *Biotechnol Bioeng* 69, 140-9.
51. Mayer, C., Zechel, D. L., Reid, S. P., Warren, A. J. & Withers, S. G., 2000, *FEBS Letters* 466, 40-4.
52. Malet, C. & Planas, A., 1998, *FEBS Letters* 440, 208-12.
53. MacLeod, A. M., Tull, D., Rupitz, K., Warren, R. A. & Withers, S. G., 1996, *Biochemistry* 35, 13165-72.
54. Moracci, M., Trincone, A., Perugino, G., Ciaramella, M. & Rossi, M., 1998, *Biochemistry* 37, 17262-70.
55. Trincone, A., Perugino, G., Rossi, M. & Moracci, M., 2000, *Bioorg Med Chem Lett* 10, 365-8.
56. De Roode, B. M. PhD Thesis, Wageningen University, Wageningen, Netherlands, 2001.
57. Riva, S. EU Enzymatic Lactose Valorization, Individual Progress Report, Milano, 1999.
58. Fischer, L., Bromann, R., Kengen, S. W., de Vos, W. M. & Wagner, F., 1996, *Biotechnology N Y* 14, 88-91.
59. De Roode, B. M. et al., 2001, *Submitted for publication*.
60. Sessitsch, A., Wilson, K. J., Akkermans, A. D. & de Vos, W. M., 1996, *Appl Environ Microbiol* 62, 4191-4.
61. Altamirano, M. M., Blackburn, J. M., Aguayo, C. & Fersht, A., 2000, *Nature* 403, 617-622.

THE XYLOGLUCAN-CELLULOSE NETWORK OF PLANT CELL WALLS: A PROTOTYPE FOR THE CHEMOENZYMATIC PREPARATION OF NOVEL POLYSACCHARIDE COMPOSITES

W. S. York, M. Pauly, Q. Qin, Z. Jia, J.P. Simon, P. Albersheim, and A.G. Darvill

The University of Georgia
Complex Carbohydrate Research Center
220 Riverbend Road
Athens GA 30602

1 INTRODUCTION

A fundamental problem in plant physiology is understanding the mechanisms by which plant cells, initially formed in the meristem, undergo controlled, oriented expansion to form mature cells with specialized morphological forms. The primary cell wall plays a key role in this process by resisting the high osmotic pressure of the plant cell (up to 1 MPa),[1] yet expanding in a highly controlled and oriented fashion. In the primary cell walls of most higher plants, the major load-bearing structure is the xyloglucan-cellulose network.[2,3] This network has been visualized in scanning electron micrographs of rapidly frozen, deep etched, rotary shadowed specimens of primary cell walls, where microfibrils appear to be cross-linked by amorphous tethers 20-40 nm in length.[4] Results obtained by sequential extraction of cell walls are consistent with identifying the crosslinks as xyloglucans.[4,5] It is clear that the controlled, oriented expansion of the xyloglucan-cellulose network is under the dynamic control of the cell itself, and involves the enzyme-catalyzed modification of its component polysaccharides as well as the biosynthesis of new polysaccharides. Nevertheless, it is likely that a better understanding of the fundamental mechanisms that contribute to the self assembly and growth of this network will improve our ability to engineer novel polysaccharide composites with valuable physical properties.

The assembly and development of this topologically complex network is likely to require both spontaneous and enzyme-catalyzed processes. For example, xyloglucan, which is quite soluble in water, spontaneously binds to the surfaces of cellulose microfibrils.[6] Thus, xyloglucan, which is synthesized in the Golgi apparatus,[7,8] can be exported in soluble form to the apoplast, where it interacts with cellulose, which is synthesized at the cell surface.[9] However, cellulose microfibrils must move relative to each other as the wall expands, and new microfibrils must be inserted into the network in order to it to maintain its structural integrity.[2] This process is likely to be mediated (in part) by a family of apoplastic enzymes, proposed by Albersheim[2] and independently discovered by Fry,[10] who called them xyloglucan endotransglycosylases (XETs), and Nishitani,[11] who called them endoxyloglucan transferases (EXTs). These enzymes catalyze the cleavage and religation of xyloglucan in the developing primary cell wall, which is likely to be a key process required for the incorporation of new material into the xyloglucan-cellulose network and for the reorganization of the network as cell morphology changes.[12-16] In addition, we have found evidence indicating that at lease two other enzymes catalyze the covalent modification of xyloglucan in the primary cell wall during growth.[17]

2 THE CHEMICAL STRUCTURE OF XYLOGLUCAN

Understanding the assembly, growth, and development of the xyloglucan-cellulose network at a molecular level first requires knowledge of the chemical structure of its components. Cellulose has a relatively simple chemical structure, consisting of long chains of β-(1→4)-linked glucosyl residues.[9] Xyloglucan (**Figure 1**) has a cellulosic backbone, and in the xyloglucans produced by most plants, ~75% of the β-(1→4)-linked glucosyl residues in the backbone are substituted with α-D-xylosyl residues at *O*6.[18,19] Many of the xylosyl residues are themselves substituted at *O*2 with β-D-Gal*p* or α-L-Fuc*p*-(1→2)-β-D-Gal*p* moieties, extending the sidechain. However, members of the plant family *Solanaceae* produce xyloglucans in which only ~50% of the backbone residues bear xylosyl residues at *O*6 and ~25% of the backbone residues bear *O*-acetyl substituents at *O*6.[20,21] The Solanaceous xyloglucans, which lack fucosyl residues, are also distinguished by having α-L-Ara*f* substituents at O2 of the α-xylosyl residues. The primary cell walls of grasses contain only small amounts of xyloglucan,[3] which is typically less branched than the xyloglucan from most plant species. Nevertheless, most plant cell walls contain xyloglucans in which 3 out of 4 backbone glucosyl residues are branched due to substitution with either an α-xylosyl residue or an *O*-acetyl group at *O*6.[21,22] The unbranched backbone residues, which are susceptible to attack by a wide range of microbial and plant endoglucanases, are regularly spaced, so treatment of xyloglucan with endoglucanase generates a well-defined series of xyloglucan subunit oligosaccharides, most of which are built around a cellotetraose core.[18-24]

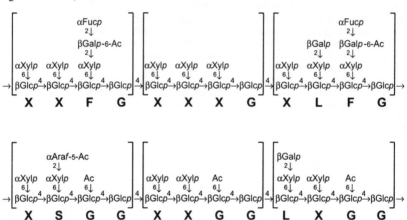

Figure 1 *Generalized structures of xyloglucans in the primary cell walls of higher plants. A typical xyloglucan produced by most higher plants is shown in the Top Panel, and a typical xyloglucan produced by Solanaceous plants is shown in the Bottom Panel. Beneath each sequence is shown the commonly used nomenclature for xyloglucan structure.*[25] *Oligosaccharides subunits generated by endoglucanase-treatment of each structure are indicated by the brackets*

If each glucosyl residue in the backbone, along with all of its pendant side chains, is defined as a primary substructure of the polymer, then the polymer can be defined as a linear array of these substructures. This is the basis for the nomenclature[25] commonly used to specify xyloglucan structures, in which each β-D-Glc*p* residue in the backbone is represented by a single uppercase letter that indicates its glycosyl substitution pattern. The chemical structure of a typical oligosaccharide subunit of the xyloglucan, composed of four of these primary substructures, is then unambiguously specified by a sequence of four uppercase letters (**Figure 1**). For example, the most common fucosylated xyloglucan subunit is represented by the sequence XXFG.

2.1 Spectroscopic Analysis of Xyloglucan Structure

The structural characterization of a xyloglucan is greatly facilitated by its endoglucanase-catalyzed fragmentation into well-defined subunit oligosaccharides. Normally, we treat the oligosaccharide fragments with sodium borohydride, converting them into the corresponding oligoglycosyl alditols. These derivatives are advantageous in that they adopt a single form, in contrast to the native oligosaccharides, whose reducing glucose residues can adopt either the α-configuration or β-configuration, which are in dynamic equilibrium. The oligoglycosyl alditols thus are much easier to isolate than the corresponding reducing oligosaccharides and have simpler NMR spectra.[18-24] We have purified more than 50 xyloglucan oligoglycosyl alditols and used multi-dimensional, high resolution NMR, in combination with tandem mass spectrometry, to determine their chemical structures. The resulting large body of spectroscopic data has made it possible to deduce many correlations between specific structural features of the oligoglycosyl alditols and the chemical shifts of diagnostic resonances in their ^{1}H-NMR spectra.[18-24] The chemical shifts of these diagnostic resonances are, to a first approximation, a function of a few, well-defined structural parameters: (*a*) the identity proton (*e.g.*, H1, H2, etc.) giving rise to the resonance. (*b*) the identity of the sugar residue containing that proton (*e.g.*, β-D-Glc*p*), including its anomeric configuration and linkage. (*c*) The identity of the primary substructure (*e.g.*, X, L, F, S, G, etc., defined above) containing the sugar residue. (*d*) The molecular environment of the primary substructure containing the nucleus, including end effects arising from its proximity to the non-reducing or alditol end of the oligomer and the presence of other side chains in the immediate vicinity. Identification of the diagnostic resonances in the ^{1}H-NMR spectra can thus provide, in principle, all of the information required to rapidly and unambiguously determine the complete glycosyl sequence of a xyloglucan oligoglycosyl alditol, simply by inspection of its 1-D ^{1}H-NMR spectrum. We have developed a database <http://www.ccrc.uga.edu/web/specdb/specdbframe.html> that provides access to the diagnostic chemical shift data we have compiled for xyloglucan oligoglycosyl alditols over the last several years.

2.2 Chromatographic Analysis of Xyloglucan Structure

Although NMR spectroscopy is a very powerful tool for determining the chemical structures of xyloglucan oligosaccharides, it is not an intrinsically sensitive method. Furthermore, NMR does not readily yield complete, unambiguous structural information for each component of a complex mixture of closely related oligosaccharides, such as those generated by endoglucanase treatment of xyloglucans. Therefore, we have developed sensitive methods to perform quantitative analysis of the complex mixtures of xyloglucan oligosaccharides.[26] This approach, described below, has made it possible to identify structural differences associated with the molecular environment within the xyloglucan-cellulose network,[27] or with differences in the developmental stage of individual plant tissues.[17]

The chromatographic method[26] we developed for analysis of xyloglucan oligosaccharides is based on the conversion of reducing oligosaccharides to their *p*-nitrobenzyloxyaminoalditol (PNB) derivatives, which are generated by reductive amination with O-(*p*-nitrobenzyl)hydroxylamine and sodium cyanoborohydride at low pH.[26] The derivatives, which are formed in yields exceeding 95%, are separated by reversed-phase HPLC and readily detected by their U.V.-absorbance at 275 nm. Individual xyloglucan oligosaccharides, including isomers that differ only in their O-acetylation pattern,[28] are identified by comparison to authentic standards, facilitated by reference to our large data base of xyloglucan oligosaccharide structures. (see above.)

3 POLYMER DOMAINS IN THE XYLOGLUCAN-CELLULOSE NETWORK

We have applied the chromatographic analysis described above to oligosaccharide mixtures prepared by a sequential extraction procedure,[27] in which purified, enzymatically depectinated, primary cell walls are treated, in turn, with a xyloglucan-specific *endo*-1,4-glucanase (kindly provided by Novozymes A/S),[29] 4N KOH, and a non-specific cellulase. Based on analysis of the oligosaccharide compositions of the extracted fractions and other data, we propose[27] that the XEG extract represents a metabolically active "enzyme-accessible domain" in the xyloglucan-cellulose network, which corresponds to the crosslinking tethers observed[4,5] in electron micrographs of primary cell walls. The fraction subsequently extracted by KOH treatment appears to be metabolically inactive, and putatively represents a xyloglucan domain that is bound to the surface of cellulose microfibrils. The final fraction is only solubilized upon enzymatic depolymerization of cellulose itself, and probably represents a xyloglucan domain that is trapped within or between cellulose microfibrils. This last domain does not appear to be covalently linked to the other two domains, which appear to be linked to each other. Comparison of these different domains provides a convenient method to detect specific metabolic processes that modify xyloglucan structure during growth and development of the cell. In addition, information obtained by this approach may provide insight into the physical mechanisms leading to the assembly of the xyloglucan-cellulose network, specifically with regard to the putative role of xyloglucan binding and intercalation in regulating the transverse dimensions of individual cellulose crystallites.

4 TOPOLOGY OF THE XYLOGLUCAN-CELLULOSE NETWORK

As mentioned above, xyloglucan is biosynthesized in soluble form in the Golgi, while cellulose is synthesized at the plasma membrane. The extracellular assembly of the xyloglucan-cellulose network occurs as the cellulose microfibril is synthesized, and is likely to involve the spontaneous binding of xyloglucan to cellulose, as well as the enzyme-catalyzed cleavage and religation of xyloglucan. Conformational energy analysis of xyloglucan suggests that xyloglucan readily adopts a "flat" or "extended" conformation in which one face of its backbone is exposed and has a shape that is complementary to the surface of the cellulose microfibril.[30] However, a "twisted" conformation has a lower calculated energy than the flat conformation, suggesting that, in solution, xyloglucan, on average, probably adopts a helical form that is not complementary to the cellulose microfibril surface.[30] This may have important consequences with respect to the topology of the xyloglucan-cellulose network (**Figure 2**). That is, the conformational change that accompanies binding of xyloglucan to the cellulose surface is accompanied by a rotation of the xyloglucan. This does not have any significant topological consequences if the xyloglucan is bound at a single point, as the unattached end of the polysaccharide should rotate freely. However, if the xyloglucan is bound at more than one site, topological complexity can arise as a result of polymer rotation. The extent and type of complexity will, of course, depend on the energetic barriers to rotations around glycosidic bonds and whether xyloglucan chains have any natural tendency to form multi-strand (*e.g.*, duplex) structures (**Figure 2**). For example, if the free energy change accompanying the formation of an antiparallel xyloglucan double helix is sufficiently lower than the energy barrier for the 360° rotation around glycosidic bonds in the xyloglucan backbone, structures such as those illustrated at the bottom of **Figure 2** could arise upon binding of xyloglucan to cellulose. The cleavage and religation of glycosidic bonds in the xyloglucan backbone catalyzed by XET are, in a topological sense, identical to the cleavage and religation reactions catalyzed by DNA topoisomerases. Thus, XET may act in the developing cell wall as a kind of polysaccharide topoisomerase. In this context, XET catalyzed-reactions might generate topologically constrained structures that would relax only upon endolytic cleavage of the xyloglucan backbone, which might be catalyzed by endoglucanases or by XET itself.

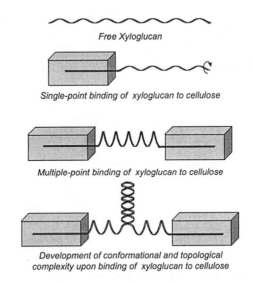

Figure 2 *Possible consequences of the binding of a xyloglucan molecule, twisted in solution, to the flat surface of a cellulose microfibril. By cleaving and religating the xyloglucan backbone, XET has the potential to generate and/or relax topological constraints within the xyloglucan-cellulose network. The characteristics of these constraints would depend on the conformational state and dynamic properties of the xyloglucan*

5 EXPERIMENTAL APPROACHES TO STUDY THE XYLOGLUCAN-CELLULOSE NETWORK

A thorough understanding of the mechanisms by which the xyloglucan-cellulose complex is assembled in the primary cell wall will require additional characterization of the conformational and dynamic properties of xyloglucan. Our initial approach to this problem is to calculate molecular dynamics trajectories for model xyloglucan oligoglycosides, and to evaluate the validity of the trajectories in light of NMR analysis of the actual model compounds. We have designed and prepared several model oligosaccharides, keeping in mind the necessity of minimizing end effects so as to more realistically mimic the molecular environments of individual oligosaccharide subunits within the xyloglucan polymer. Enzymatic techniques were used extensively to prepare the oligoglycosides using polymeric xyloglucan as the starting material.

5.1 Endoglucanase-Catalyzed Transglycosylation Reactions

Native oligosaccharides are not good model compounds for our conformational analyses, as their reducing glucosyl residues are in dynamic equilibrium, interconverting between the α- and β-anomeric forms. Thus, they do not faithfully mimic the molecular environment of a xyloglucan subunit within the polysaccharide. We have used transglycosylation reactions catalyzed by a family 7 endoglucanase (EG) to prepare a range of xyloglucan β-alkyl oligoglycosides.[31] In these reactions, xyloglucan polysaccharide is incubated with the enzyme in aqueous solutions containing ~20% alcohol, which acts as an acceptor substrate. The transglycosylation products (β-alkyl glycosides) are obtained in good yield (up to 70%) and are readily separated from hydrolysis products (reducing oligosaccharides) generated in these reactions. A broad range of alcohols, including several with secondary reactivities, are efficient acceptor substrates for the EG, generating alkyl glycosides that have great potential as intermediates in the synthesis of novel, polysaccharide-based products.

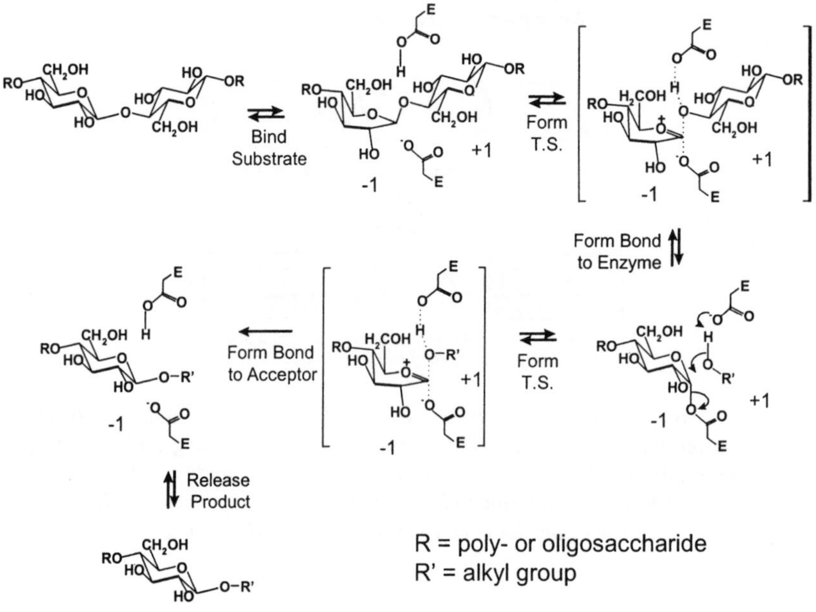

Scheme 1 *Proposed basis for the stability of β-alkyl glycoside transglycosylation products generated by family 7 EGs, based on the widely accepted catalytic mechanism for these configuration-retaining enzymes* [32]

We have found that the β-glycoside products generated by these reactions are not generally good substrates for the endoglucanase.[31] This constitutes a distinct advantage of this approach, as, once the products are made, they are stable in the presence of active EG even if water is added or the alcohol is removed. This observation is consistent with the proposed configuration-retaining catalytic mechanism for family 7 EGs,[32] in which initial binding of the substrate induces conformational distortion of the glucosyl residue in the -1 subsite, situating the leaving group in the preferred axial position (**Scheme 1**). Even if the β-alkyl glycoside transglycosylation product binds to the active site, its alkyl aglycon apparently does not have sufficient or appropriate energetic interactions with the +1 subsite to induce distortion of the glucosyl residue in the -1 subsite. That is, the transition state has a significantly higher free energy than the enzyme-bound alkyl glycoside, and so formation of the glycosidic bond to the acceptor alcohol is, for all practical purposes, irreversible, as indicated by the single arrow for the penultimate step in **Scheme 1**.

5.2 Transglycosylation Reactions Catalyzed by Isoprimeverosyl Hydrolase

Another enzyme that we have used extensively to prepare and characterize xyloglucan oligoglycosides is isoprimeverosyl hydrolase (IPH), which cleaves the disaccharide isoprimeverose (6-*O*-α-D-xylopyranosyl-D-glucose) from the nonreducing ends of xyloglucan oligosaccharides. Although it is a well known component of several fungal glycosyl hydrolase preparations, including driselase (from *Irpex lacteus*, a basidiomycete),[33] and Sanzyme 1000 (*Aspergillus oryzae*),[34] IPH is difficult to purify to homogeneity. We observed that an *Aspergillus* IPH has significant transglycosylating activity, transferring isoprimeverose to the non-reducing end of xyloglucan oligosaccharides. Analysis of the transferase products indicated that a β-(1→4) linkage is formed when XXG is the acceptor substrate (generating XXXG), but that both β-(1→4) and β-(1→3) linkages are formed when XXXG is the acceptor substrate (generating XXXXG and X³XXXG, where the superscript 3 indicates an unusual β-(1→3) linkage). This suggests that when XXXG acts as an acceptor substrate, it can bind to the active site in two different orientations. The mechanism of IPH-catalyzed transglycosylation reaction may thus have similarities to the mechanisms of glycosyl transfer reactions leading to mixed-linkage glucans in cereals.

Oligosaccharides (*i.e.*, XXXXG) with five glucosyl residues in the backbone have been reported[35] in EG-digests of the xyloglucan from *Hymenaea courbaril* L. cotyledons. The detection of XXXXG in cotyledons is quite unexpected, as primary xyloglucan synthesis in the Golgi almost always produces a polymer comprised of subunits that have four glucosyl residues in the main chain.[22] Thus, although no β-(1→3) linkages were reported in the *H. courbaril* xyloglucan,[35] it is possible that the observed XXXXG was generated by an IPH with transglycosylating activity.

6 CONCLUSIONS

Assembly of the topologically complex xyloglucan-cellulose network in plant cell walls involves spontaneous processes, such as the adhesion of soluble xyloglucan to the surface of cellulose microfibrils, and enzyme-catalyzed processes, such as the cleavage and religation of the xyloglucan backbone. Several of the enzymes involved in the assembly and restructuring of the xyloglucan-cellulose have been isolated and cloned. These enzymes can be used, together with enzymes from other sources and chemical methods, to prepare novel oligomeric and polymeric materials.

7 ACKNOWLEDGMENTS

This work was supported by US Department of Energy (DOE) grant DE-FG05-93ER20220, the DOE-funded (DE-FG05-93ER20097) Center for Plant and Microbial Complex Carbohydrates, and the National Science Foundation (NSF) grant MCB-9974673.

References

1.   D. J. Cosgrove, *Plant Cell*, 1997, **9**, 1031.
2.   P. Albersheim, in *Plant Biochemistry*. 3rd Edn., ed. J. Bonner, and J.E. Varner,. London, Academic Press. 1976, Chapter 9, p. 225.
3.   N. C. Carpita and D. M. Gibeaut, *Plant J.*, 1993, **3**, 1.
4.   M.C. McCann, B. Wells, and K. Roberts, *Journal of Cell Science,* 1990, **96**, 323.
5.   T. Fujino, Y, Sone, Y. Mitsuishi, and T. Itoh, *Plant Cell Physiol.,* 2000, **41**, 486.

6.   B. S. Valent and P. Albersheim, *Plant Physiol.*, 1974, **54**, 105.
7.   A. Driouich, L. Faye, and L. A. Staehelin, *Trends in Biochemical Sciences*, 1993, **18**, 210.
8.   D. A. Brummell, A. Camirand, and G. A. Maclachlan, *J. Cell Sci.*, 1990, **96**, 705.
9.   D. P. Delmer, *Annu. Rev. Plant Physiol. Plant Mol. Biol.*, 1999, **50**, 245.
10. S. C. Fry, R. C. Smith, K. F. Renwick, D. J. Martin, S. K. Hodge, and K. J. Matthews, *Biochem. J.*, 1992, **282**, 821.
11. K. Nishitani and R. Tominaga, *J. Biol. Chem.*, 1992, **267**, 21058.
12. S. C. Fry, *Annu. Rev. Plant Physiol. Plant Mol. Biol.*, 1995, **46**, 497.
13. K. Nishitani, *International Review of Cytology*, 1997, **173**, 157.
14. J. Braam, *Trends Plant Sci.*, 1999, **4**, 361.
15. D. J. Cosgrove, *Annu. Rev. Plant Physiol. Plant Mol. Biol.*, 1999, **50**, 391.
16. J. K. C. Rose and A. B. Bennett, *Trends Plant Sci.*, 1999, **4**, 176.
17. M. Pauly, Q. Qin, H. Greene, P. Albersheim, A. Darvill, and W. S. York, *Planta*, 2001, **212**, 842.
18. Kooiman, P. *Rec. Trav. Chim.*, 1961, **80**, 849.
19. W. S. York, H. van Halbeek, A. G. Darvill, and P. Albersheim, *Carbohydr. Res.* 1989, **200**, 9.
20. W. S. York, V. S. Kumar Kolli, R. Orlando, P. Albersheim, and A. G. Darvill, *Carbohydr. Res.* 1996, **285**, 99.
21. Zhongua Jia and William S. York, unpublished results.
22. J.-P. Vincken, W. S. York, G. Beldman, and A. G. Voragen, *Plant Physiol.*, 1997, **114**, 9.
23. W. S. York, G. Impallomeni, M. Hisamatsu, P. Albersheim, and A. Darvill, *Carbohydr. Res.* 1994, **267**, 79.
24. W. S. York, L. K. Harvey, R. Guillén, P. Albersheim, and A. G. Darvill, *Carbohydr. Res.* 1993, **248**, 285.
25. S. C. Fry, W. S. York, P. Albersheim, A. G. Darvill, T. Hayashi, J.-P. Joseleau, Y. Kato, E. P. Lorences, G. A. Maclachlan, M. McNeil, A. J. Mort, J. S. G. Reid, H.-U. Seitz, R. R. Selvendran, V. N. Shibaev, A. G. J. Voragen, and A. R. White, *Physiologia Plantarum*. 1993, **89**, 1.
26. M. Pauly, W. S. York, R. Guillén, P. Albersheim, and A. G. Darvill, *Carbohydr. Res.* 1995, **282**, 1.
27. M. Pauly, P. Albersheim, A. Darvill, and W. S. York, *The Plant J.*, 1999, **20**, 629.
28. W. S. York, J. E. Oates, H. van Halbeek, A. G. Darvill, P. Albersheim, P. R. Tiller, and A. Dell, *Carbohydr. Res.* 1988, **173**, 113.
29. M. Pauly, L. N. Andersen, S. Kaupinen, L. V. Kofod, W. S. York, P. Albersheim, and A. G. Darvill, *Glycobiology*, 1999, **9**, 93.
30. S. Levy, W. S. York, R. Struike-Prill, B. Meyer, and L. A. Staehelin, *The Plant J.* 1991, **1**, 195.
31. W. S. York and R. Hawkins, *Glycobiology*, 2000, **10**, 193.
32. G. Sulzenbacher, H. Driguez, B. Henrissat, M. Schülein, M., and G. J. Davies, *Biochemistry* 1996, **35**, 15280.
33. E. P. Lorences and S. C. Fry, *Carbohydr. Res.*, 1994, **263**, 285.
34. Y. Kato, J. Matsushita, T. Kubodera, and K. Matsuda, *J. Biochem.*, 1985, **97**, 801.
35. M. S. Buckeridge, H. J. Crombie, C. J. M. Mendes, J. S. G. Reid, M. J. Gidley, and C. C. J. Vieira, *Carbohydr. Res.*, 1997, **303**, 233.

6 Enzymology of Plant Cell Wall Carbohydrates

CELLULOSE SYNTHESIS AND ENGINEERING IN PLANTS

S. R. Turner, N. G. Taylor and P. Szyjanowicz

University of Manchester
School of Biological Science
3.614 Stopford Building
Oxford Road
Manchester M13 9PT
United Kingdom

1 INTRODUCTION

Plant cell walls may be classified as primary or secondary cell walls. Primary cell walls are synthesised while the cell is still expanding. The orientation of cellulose microfibrils within the primary cell wall is thought to control the direction of cell expansion. Consequently, cellulose within the plant cell wall has a key role in controlling cell shape and hence plant morphology. In contrast, secondary cell walls are laid down once the cell has attained its final shape. These secondary cell walls are responsible for the mechanical strength of plant material.

Pressure to improve raw material for the pulp and paper industry as well as improved forage crops have yielded some success. These successes, however, are largely based on generating material with enhanced processing properties using genetic engineering to reduce lignin content. Whilst in some circumstances decreased lignin content may have indirectly increased cellulose content[1], efforts to increase yield or improve raw material properties are hampered by a relatively poor understanding of how cellulose is synthesised in higher plants. Efforts to understand cellulose synthesis in higher plants have yielded few results due to an inability to purify the cellulose synthase complex, whilst retaining high enzyme activity.

1.1 Identification of genes involved in cellulose synthesis in plants

Until relatively recently, no genes for any of the subunits of the higher plant cellulose synthase subunit had been cloned. Pear et al.[2] were the first to describe a clone from cotton (now described as a member of the CesA gene family) which showed homology to the catalytic subunit of bacterial cellulose synthases, including those from *Acetobactor xylinum* and *Agrobacterium tumefaciens*. These proteins all contain several conserved sequences indicative of a processive glucosyl transferase. Further proof that a member of this gene family encodes the catalytic subunit of the higher plant cellulose synthases complex was provided by studies on a temperature sensitive mutant of *Arabidopsis* (*rsw1*).[3] At 31^{0}C, *rsw1* plants die at an early stage of development and have only half the cellulose content of wild type. *rsw1* has a mutation in a member of the CesA family of genes.[3]

Analysis of the completed *Arabidopsis* genome suggests that it contains a superfamily containing more than 50 genes showing homology to bacterial cellulose synthases (http://cellwall.stanford.edu/cesa/index.shtml). The CesA genes form a clear subfamily. There are at least 10 members of the CesA gene family in *Arabidopsis*. The role of these different CesA family members is an area of intense interest.

1.2 Structure of the plant cellulose synthase complex

Much of the information on the structure of the higher plant cellulose synthase complex has come from scanning electron microscopy of freeze fractured plasma membranes. These studies have revealed the existence of rosettes made up of six 'globules' embedded in the plasma membrane. Various lines of evidence suggested that these rosettes are likely to be the site of cellulose synthesis.[4] Studies on the temperature sensitive mutant *rsw1* provide further evidence that these structure are indeed the sites of cellulose synthesis. At the restrictive temperature *rsw1* plants exhibit reduced cellulose in the primary cell wall, the breakdown of the organisation of rosettes to give disorganised globules.[3] This observation indicates an essential role for CesA in both the catalysis of β(1-4) linked glucose and the organisation of the cellulose synthase complex.

Delmer and Amor[4] have suggested that each globule of the rosette contain six subunits of each polypeptide required to synthesise cellulose. According to this model each rosette would be a '36mer', simultaneously synthesising 36 chains of cellulose. Consequently, this rosette structure may contain up to 36 copies of several different proteins, making it a large multi-protein complex. Measurement using with solid state NMR,[5] however, have suggested that cellulose is synthesised initially as an 'elementary fibril', which is composed of approximately 18 chains. To date it is still unclear exactly how many β(1-4) glucose chains are synthesised by a single rosette and whether a rosette synthesises one or more elementary fibrils. Many of these questions may potentially be answered with a proper understanding of how the different CesA proteins are organised within the rosette.

1.3 Mutants with defective cellulose synthesis in the secondary cell wall

The many advantages of *Arabidopsis* as a model for molecular genetic analysis are well documented. In addition, rates of secondary cell wall synthesis are high in stems of the appropriate age. Whilst cellulose constitutes only a small percentage of seedlings, up to 35% of the ethanol-insoluble fraction of mature stems is cellulose.[6] Consequently developing stems are an excellent source of starting material for any biochemical analysis. Most importantly it is possible to isolate very severe mutations. We have previously isolated *Arabidopsis irregular xylem* (*irx*) mutants that exhibit defects in their secondary cell wall. To date 5 complementation groups have been identified (table 1). One complementation group is defective in lignin deposition, whilst the remaining 4 synthesise little or no cellulose in the secondary cell wall[6]. For example, the total cellulose content in stem segments of *irx3* plants is decreased more than 5-fold (the reduction in secondary cell wall cellulose is likely even greater), but despite this the plants remain relatively healthy[6]. Both light and electron microscopy show that *irx3* plants have thin, uneven, darkly staining secondary walls.

Table 1 *Irregular xylem mutants of* Arabidopsis thaliana

Complementation group	No. alleles	Defect
irx1	4	Cellulose deficient
irx2	2	Cellulose deficient
irx3	2	Cellulose deficient
irx5	3	Cellulose deficient
irx4	1	Lignin deficient
unassigned	3	Unknown

2 INCREASING CELLULOSE CONTENT BY ALTERING CARBON FLUX

There are few documented examples of increasing cellulose content in plants. One example comes from work on sucrose metabolism in cotton fibres. Overexpression of sucrose phosphate synthase (SPS) has resulted in increased yield in cotton fibres. These fibres have thicker secondary cell walls that contain more cellulose and increased mechanical strength (C. Haigler pers. comm.). The reason for this increase in cellulose remains under investigation. It is likely, however, to stem from the association of sucrose synthase with the cellulose synthase complex and the coupling of sucrose breakdown to form UDP-glucose and the polymerisation of cellulose. Whilst sucrose may be supplied from other parts of the plant, two reasons suggest that it may be necessary to make cellulose within the cell. Firstly, product of sucrose synthase yields fructose that must be recycled. Secondly, sucrose may be broken down within the cell by the actions of invertase. Upregulating SPS is clearly an excellent target for increasing flux into sucrose within the cell.

3 COMPONENTS OF THE PLANT CELLULOSE SYNTHASE COMPLEX

Whilst purification of the plant cellulose synthase complex with high activity remains elusive, significant progress in identifying components of the complex has been made with genetic analysis. Identification of a number of mutants, mostly notably in Arabidopsis has lead to rapid progress in the field.

3.1 Identification of the Presumed Catalytic Subunit

The completion of the Arabidopsis genome has greatly changed many aspects of plant biology. One important aspect is the way in which it has greatly facilitated map based cloning. Even in the years immediately prior to the completion of the sequence, the available sequence information greatly aided map based cloning.

3.1.1 Cloning IRX1 and IRX3. The *irx3* mutant was initially mapped to the top arm of chromosome V. Analysis of a large number of ESTs showing homology to bacterial cellulose synthases[7] revealed that one of these ESTs (75G11) mapped to a region close to *irx3*. Subsequent analysis demonstrated that the *irx3* mutation was indeed caused by a mutation in this gene.[8] A systematic nomenclature has been suggested by Delmer.[9] Using an updated form of this nomenclature (http://cellwall.stanford.edu/cesa/index.shtml) the IRX3 gene corresponds to AtCesA7.

Mapping of *irx1* revealed that the mutation was located on the top arm of chromosome 4. Completion of the genome sequence in this region revealed the presence of another member of the CesA gene family (AtCesA8). Confirmation that *irx1* was indeed caused by a mutation in this gene was demonstration by complemenation.[10]

The secondary cell walls of the tracheary elements from both of these mutants have characteristic even, thin, dark-staining secondary cell walls.[10] Examination of the tracheary elements of the xylem using light and transmission electron microscopy indicated that both the *irx1* and *irx3* mutations appear to give rise to an identical phenotype. This clearly suggests that both gene products are required for cellulose synthesis within the same cell type.

3.1.2 Analysis of the Interaction between IRX1 and IRX3. An epitope tag has been inserted into the *IRX3* gene. Initial experiments utilised an RGSHis tag, which contains a run of 6 histidines for use in immobilised metal affinity chromatography as well as a recognition site for a monoclonal antibody. Insertion of this tag close to the N-terminus results in a protein that may be recognised using the monoclonal antibody (fig. 1). Insertion of this tag does not disrupt the way in which the protein functions since the recombinant protein still complements the *irx3* mutation.[10] At only 9 amino acids the RGSHis tag is comparatively small, recent work has demonstrated that it is possible to add GFP at the same site and still retain normal activity (N. Taylor and S. Turner unpublished). In addition, we have raised highly specific antibodies to the variable regions of both IRX1 and IRX3 as well as to the constant region of IRX3 (fig. 1).

The epitope-tagged IRX3 protein was solubilised in Triton X-100 and incubated with a metal affinity resin. Using an IRX1 specific polyclonal antibody it was possible to demonstrate that a similar proportion of IRX1 was co-precipitated with the IRX3 protein.[10] Precipitation of IRX3 is completely dependent upon the hexa-histidine tag, whilst other plasma membrane markers did not co-precipitate. These results demonstrate that IRX1 and IRX3 directly interact and that they are likely to be part of the same enzyme complex.[10]

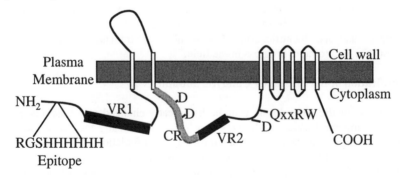

Figure 1 *Diagram showing the predicted topology of the IRX3 protein. The approximate position of the two variable regions (VR1 and VR2) and a highly conserved region (CR) are indicated. Four motifs (three aspartate residues and QxxRW) indicative of processive glycosyltransferases are also shown. Antibodies described in the text were raised to polypeptides from VR1.*

3.1.3 Localisation of IRX3 and IRX1. Tissue printing is a convenient means of examining the localisation of proteins within the stem. Using successive prints from the cut surface of a mature inflorescence stem it is clear that IRX3 and IRX1 have a very similar distribution. Both proteins localise to the xylem and to the cells of the interfascicular region.

3.1.4 Characterisation of IRX5. Recently, an allele of another cellulose deficient complementation group, *irx5* was identified as being caused by an insertion of a Ds transposon. Analysis of the flanking sequences revealed that *irx5* was caused by an insertion into another member of the CesA gene family (AtCesA4) (A Huttly, pers comm.). Using similar methods to those described above for IRX1 and IRX3 it has been shown that IRX5 is part of a complex comprising IRX1, IRX3 and IRX5 (N Taylor and S. Turner unpublished). Since the tracheary elements of the xylem of *irx5* exhibit an identical phenotype to both *irx1* and *irx3* it is clear that all three gene products are essential for cellulose synthesis in these cell types.

3.1.5 Are three CesA genes required to make cellulose in all cell types? Our data suggests that at least three members of the *Arabidopsis* CesA gene family are required to make cellulose in xylem cells (Table 2). Whilst the phenotype of *irx3* mutant plants show dramatic alterations in cellulose content in both the xylem and interfascicular region the phenotype of *irx1* plants exhibit a much less dramatic affect on the interfascicular cells. It is unclear at present why the phenotype of *irx1* plants is less pronounced in the interfascicular region. However, there is the possibility that some functional redundancy exists and another member of the gene family may be able to compensate for the absence of IRX1 function. In addition, IRX1, IRX3 and IRX5 are all apparently specific for secondary cell wall cellulose biosynthesis. Mutations in *RSW1* (AtCesA1)[3] or *PRC1*[11] (AtCesA6) appear to affect the primary cell wall. In addition, *ISOXABEN RESISTANCE1* (AtCesA3) may also be a cellulose synthase involved in primary cell wall biosynthesis. Consequently, it is possible that two non redundant groups of three CesA genes are required to make cellulose in the primary (AtCesA1, 3 and 6) or secondary cell wall (AtCesA4, 7 and 8). Many question about how these rosettes are organised and why so many different family members are required awaits further study.

3.2 Are Endoglucanases Required for Cellulose Synthesis

The first description of the gene *KORRIGAN (KOR)* came from work on a mutant that was described as having reduced cell elongation. The KOR gene product showed clear homology to a β(1-4) endoglucanase and appeared to be membrane bound[13]. A more severe allele was later described as being a cell plate specific endoglucanase essential for normal cell division in plants.[14] Further alleles of this mutation have also been described with cell wall defects.[15,16,17] Whilst *kor* alleles may show alterations in pectic composition,[18] the primary defect appears to be caused by a decrease in cellulose deposition.[15,16,17] This is further confirmed by making double mutants between weak *korrigan* alleles and weak *rsw1* alleles. These double mutants show a larger decrease in cellulose content than either of the single mutants,[16,17] strongly suggesting that both genes are required for cellulose deposition.

More recently it has also become clear that *irx2* is also allelic with *kor* (P. Szyjanowicz and S. Turner unpublished). Interestingly, *irx2* plants exhibit no apparent radial swelling or growth defects characteristic of other *kor* alleles. The *irx2* mutations appear to primarily affect secondary cell wall synthesis, even though one allele is caused by a mutation in a highly conserved amino acid.

4 PROPECTS FOR CELLULOSE ENGINEERING IN PLANTS

Identification of the essential components for cellulose synthesis in plants remains the first step in understanding how cellulose is regulated. Whether by overexpressing the 4 components identified to date will be sufficient to give increased cellulose synthesis remains to be seen. It is likely that a number of additional factors will also be required. Until regulatory genes are discovered it remains unlikely that this approach will lead to increased cellulose production in the near future.

Table 2 *Known Cellulose Deficient Mutants of* Arabidopsis

Mutation	Gene affected	Wall affected	Reference
irregular xylem1 (irx1)	AtCesA8	Secondary	10
irregular xylem3 (irx3)	AtCesA7	Secondary	8
irregular xylem5 (irx5)	AtCesA4	Secondary	Taylor and Turner unpub.
radial swelling1 (rsw1)	AtCesA1	Primary	3
isoxaben resistant (ixr1)	AtCesA3	Primary	Scheible and Somerville unpub.
isoxaben resistant (ixr2)	AtCesA6	Primary	
procuste (prc1)			11
Quill			12
Korrigan	Endo-		13
radial swelling2 (rsw2)	glucanase		15,17
altered cell wall1 (acw1)			16
irregular xylem2 (irx2)			Szyjanowicz and Turner unpub.

References

1. W-J Hu, S.A. Harding, J. Lung, J. Popko, J. Ralph, D. Stokke, C-J Tsai and V.L. Chiang, Nat. Biotech. 1999, **17**, 808-812
2. J.P. Pear, Y. Kawagoe, W.E. Schrenkengost, D.P. Delmer and D.M. Stalker, Proc. Natl. Acad. Sci, USA 1996, **93**, 12637-12642.
3. T. Arioli, L. Peng, A.S. Betzner, J. Burn, W. Wittke, W. Herth, C. Camilleri H. Hofte J.Plazinski, R. Birch and R Williamson, Science 1998, **279**, 717-720.
4. D.P. Delmer and Y.Amor, Plant Cell 1995, **7**, 987-1000.
5. M.A. Ha, D.C. Apperley, B.W. Evans, M. Huxham, W.G. Jardine, R.J. Vietor, D. Reis, B. Vian and M.C. Jarvis, Plant Journal 1998,**16**, 183-190.
6. S.R. Turner and C.R. Somerville, Plant Cell 1997, **9**, 689-701.
7. S. Cutler, and C.R. Somerville, Curr Biol 1997, **7**, R108-R111.

8. N.G. Taylor W.-R. Schieble, S. Cutler, C.R. Somerville and S.R. Turner, Plant Cell, 1999, **9**, 689-701.
9. D.P. Delmer, Ann. Rev. Plant Physiol. Plant Mol. Biol. 1999, **50**, 245-276.
10. N.G. Taylor, S. Laurie and S.R. Turner, Plant Cell 2000, **12**, 2529-2539.
11. M. Fagard. T. Desnos, T. Desprez, F. Goubet, G. Refegier, G. Mouille, M. McCann, C. Rayon, S. Vernhettes and H. Hofte, Plant Cell, 2000, **12**, 2409-2423.
12. M.-T. Hauser, A. Morikami, and P.N. Benfey, Development 1995, **121**, 1237-1252
13. F. Nicol, I. His, A. Jauneau, S. Vernhette S, H. Canut and H. Hofte, EMBO J. 1998, **17**, 5563-5576
14. J. Zuo, Niu QW, Nishizawa N., Y. Wu, B. Kost and N.H. Chua, Plant Cell 2000. **12**, 1137-1152
15. L. Peng, C.H. Hocart, J.W. Redmond and R.E. Williamson, Planta 2000, **211**, 406-414
16. S. Sato, T Kato, K. Kakegawa, T. Ishii, Y-G, Liu, T. Awano, K. Takabe, Y. Nishiyama, S. Kuga, S. Sato, Y. Nakamura S. Tabata and D. Shibata, Plant Cell Physiol. 2001 **42**, 251-263
17. D.R. Lane, A. Wiedemeirer, L. Peng, H. Hofte, S. Vernhettes, T. Desprez, C.H. Hocart, R.J. Birch T.I. Baskin, J.E. Burn, T. Arioli, A.S. Betzner and R.E. Williamson, Plant Physiol. 2001, **126**, 278-288
18. I. His, A. Driouch, F. Nicol, A. Jauneau and H. Hofte, 2001, Planta **212**, 348-358

STUDIES ON PLANT INHIBITORS OF PECTIN MODIFYING ENZYMES: POLYGALACTURONASE-INHIBITING PROTEIN (PGIP) AND PECTIN METHYLESTERASE INHIBITOR (PMEI)

B. Mattei[1], A. Raiola[1], C. Caprari[1], L. Federici[2], D. Bellincampi[1], G. De Lorenzo[1], F.Cervone[1], A. Giovane[3], L. Camardella[4]

[1]Department of Plant Biology
[2]Department of Biochemical Sciences, University of Roma "La Sapienza", Roma, Italy ;
[3]Department of Biochemistry and Biophysics, II University of Napoli, Italy
[4]Institute of Protein Biochemistry, C.N.R.,Napoli,Italy

1 INTRODUCTION

Pectin of the plant cell wall is either degraded during microbial attacks or remodeled during the various phases of growth and development.

We are studying two protein inhibitors of pectic enzymes: polygalacturonase inhibiting protein (PGIP) and pectin methylesterase inhibitor (PMEI). PGIP is specific for fungal polygalacturonases (PGs) and is involved in defense against phytopathogenic fungi by allowing the accumulation of elicitor active oligogalacturonides (OGs). PMEI is active against plant pectin methylesterase (PME) and likely involved in the regulation of the degree of methylation of the cell wall pectin during plant growth and development.

2 POLYGALACTURONASE INHIBITING PROTEINS (PGIP)

The occurrence of PGIPs has been reported in a variety of dicotyledonous plants,[1] as well as in the pectin-rich monocotyledonous plants onion and leek.[2,3] The genes encoding PGIPs are organized in families, where the different members of each family encode proteins with nearly identical characteristics but different specificity and regulation. In bean cells transcripts accumulate following addition of elicitors such as OGs and fungal glucan and in response to wounding or treatment with salicylic acid;[4] rapid accumulation of mRNA correlates with the appearance of the hypersensitive response in incompatible interactions.[5]

PGIPs belong to a the family of leucine-rich repeat (LRR) proteins. In plants, LRR proteins play a relevant role in both development and defense, where specificity of recognition is a fundamental prerequisite.

The mature PGIPs are characterised by the presence of 10 repeats, each derived from modifications of a 24-amino acid leucine–rich peptide. In *P. vulgaris* PGIP, the domain from amino acid 69 to 326 can be divided into a set of 10.5 tandemly repeating units, each derived from modifications of a 24-amino acid peptide; the repeat element matches the extracytoplasmic consensus LxxLxxLxxLxLxxNxLxGxIPxx (Figure 1).[6]

```
A    10   MSSSLSIILVILVSLRTAHS

B    30   ELCNPQDKQALLQIKKDLGNPTTLSSW
            LPTTDCCNRTWL

                          XXLXLXX
                           β    β
                         strand turn
C    69    GVLCDTDTQT  YRVNNLDLSG  LNLPKP
     95    YPIPSSLANL  PYLNFLYIGG  INNLV
     120   GPIPPAIAKL  TQLHYLYITH  TNVS
     144   GAIPDFLSQI  KTLVTLDFSY  NALS
     168   GTLPPSISSL  PNLVGITFDG  NRIS
     192   GAIPDSYGSFSKLFTSMTISR   NRLT
     217   GKIPPTFANL   NLAFVDLSR  NMLE
     240   GDASVLFGSD  KNTQKIHLAK  NSLA
     264   FDLGKVGLS   KNLNGLDLRN  NRIY
     287   GTLPQGLTQL  KFLHSLNVSF  NNLC
     311   GEIPQG GNL  QR  ⌐ FDVSA ⌐

D    327   YANNKCLCGSPLPACT

consensus
sequence   G.IP..L..L.KNI..LDLS..NNL
```

Figure 1 *PGIP-2 LRR structure. A, signal peptide; B, presumed N-terminus of the mature protein; C, 10.5 LRRs and D, C-terminus. The box indicates the area of the protein predicted to form the β-sheet/β-turn structural motifs*

2.1 The recognition abilities of PGIPs

Against the many PGs produced by fungi, plants have evolved different PGIPs with specific recognition abilities. PGIPs are typically effective against fungal PGs and ineffective against other pectic enzymes either from microbial or plant origin.[7]

PGIPs apparently inhibit a range of PGs with *endo/exo* mode of substrate degradation, but discriminate between PGs with classic *endo-* pattern of cleavage.

Not only do PGIPs from different plant sources differ in their inhibitory activities, but also PGIPs from a single plant source inhibit, with different strength, PGs from different fungi or different PGs from the same fungus. For example the inhibitory activity of bulk bean PGIP is a composite of activity of several PGIPs. Crude preparations of bean PGIP are effective against PGs from *Aspergillus niger*, *Botrytis cinerea* and *Fusarium moniliforme*. Interestingly, bean PGIP is significantly more effective against PG of *Colletotrichum lindemuthianum* than against PGs of the related non pathogen *Colletotrichum lagenarium*, suggesting that compatibility determines a selection pressure for more efficient PGIPs that can better counteract fungal infection.

Two bean PGIPs with nearly identical biochemical characteristics but with distinct inhibitory activity have been separated by differential affinity chromatography: one of them inhibits PG from *A. niger*, but not PG from *F. moniliforme*, while the other inhibits both enzymes (Figure 2).[8]

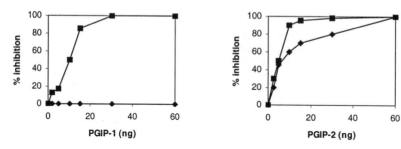

Figure 2 *Inhibition curves of* Aspergillus niger *PG* ■ *and* Fusarium moniliforme ◆ *from increasing amounts of purified PGIP*

2.2 The structural basis of PG-PGIP interaction

A significant advance in understanding the structural basis of LRR-mediated interactions comes from studies of the porcine RNAse inhibitor (PRI). PRI consists of β-strand/β-turn structural units, arranged for a parallel beta-sheet so that the protein acquires an unusual horseshoe-like shape[9].

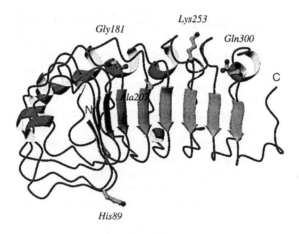

Figure 3 *The fold of modelled PGIP-1 . The position of the five amino acid differences between PGIP-1 and PGIP-2 that lie in the LRR domain are shown*

The repeated β-strand/β-turn structure is determined by the presence in each LRR module of the motif xxLxLxx, where the leucine residues form a hydrophobic core, while the sidechains of the amino acids flanking the leucines are exposed to the solvent and interact with the ligand.[10] Modeling studies suggest that, like PRI, PGIP exhibits a parallel stacking of β-strand/β-turns forming a solvent-exposed surface; the protein assumes a arch-shaped protein fold resembling that of the β-helical structure (Figure 3).[11]

A detailed analysis of the secondary structure of bean PGIP1 by far-UV CD and infrared spectroscopy coupled to constrained prediction methods indicates the presence of 12 α and 12 β secondary structure segments. The protein consists of three domains, namely the central LRR region and two cysteine-rich flaking domains. The four cysteine residues at the N-terminal end form two disulfide bridges, and at the C-terminal end four similarly spaced cysteine residues in a 27-amino acid stretch are also bridged by disulfide bonds, indicating that these linkages are important to maintain structures of local importance, and to provide additional stabilization of the secondary and tertiary structure.

The alignment of known PGIP sequences shows that the positions of cysteine residues are highly conserved in all the PGIPs. This could reflect the presence of conserved structural features of general importance for the biological function of these LRR proteins. Two N-linked oligosaccharides are located on Asn 64 and Asn 141. The main structure resembles the typical complex plant N-glycan consisting of a core pentasaccharide β-1,2-xylosylated, carrying an α-1,3-fucose linked to the innermost N-acetylglucosamine and one outer arm N- acetylglucosamine residue (Figure 4).[12]

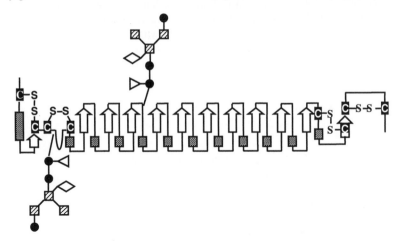

Figure 4 *Schematic draw of PGIP secondary structure elements. Arrows indicate β strands and boxes α helices. Glycan structure has been sketched as follows:*

(●) GlcNAc, (▨) Man, (▽) Fuc, (◇) Xyl.

The amino acids of PGIP which determine specificity and affinity with fungal PGs are internal to the predicted solvent-exposed β-sheet/β-turn structure. The mature proteins encoded by pgip1 and pgip2 of *P. vulgaris* differ only by 8 amino acid residues located preferentially within or contiguous to the motif xxLxLxx. The two proteins exhibit distinct specificities: PGIP1 is not able to interact with PG of *F. moniliforme* and interacts with PG of *A. niger*; PGIP2 interacts with both PGs. Single mutations of amino acid residues of PGIP2 into the corresponding residues of PGIP1 cause a loss of affinity towards *F. moniliforme* PG. On the other hand, the single mutation of a lysine at position 253 of PGIP1 into the corresponding amino acid of PGIP2, a glutamine, is sufficient to confer to the protein the capacity of interacting with *F. moniliforme* PG.[11] These data provide the evidence that variations in the solvent-exposed β-sheet/β-turn structure of PGIP have a functional significance and determine the discriminatory ability for a specific recognition of PG.

The residues of PG involved in the interaction with PGIP are also under investigation.[13] Recently, the 3-D structure of *F. moniliforme* PGs (Federici *et al*, unpublished) have been elucidated. The enzyme forms multiple contacts with the inhibitor; Lys269 and Arg 267, located inside the active side cleft and His 188, at the edge of the active site cleft, are among the residues directly involved in the formation of the complex.

3 PECTIN METHYLESTERASE INHIBITOR (PMEI)

A protein acting as a powerful inhibitor of PME was discovered in ripe kiwi (*Actinidia chinensis*) fruit.[14,15] The protein inhibits plant PME through the formation of a non-covalent 1:1 complex. The PME inhibitor (PMEI), which has been detected only in kiwi fruit, is probably synthesized as a larger protein which undergoes post-translational processing.[15] Unlike PME, PMEI is not tightly bound to cell wall and can be easily extracted from the ripe fruits at low ionic strength allowing a simple separation from the endogenous PME. At present the inhibitor has been shown to be active only on PME of plant origin; therefore a physiological role, possibly on the regulation of the fruit ripening process, can be hypothesized.

3.1 PMEI primary structure

```
                  20                        40
ENHLISEI■PKTRNPSL■LQALESDPRSASKDLKGLGQFS
                  60                        80
IDIAQASAKQTSKIIASLTNQATDPKLKGRYET■SENYAD
                 100                       120
AIDSLGQAKQFLTSGDYNSLNIYASAAFDGAGT■EDSFEG
                 140              152
PPNIPTQLHQADLKLEDLCDIVLVISNLLPGS
```

Figure 5 *Amino acid sequence of PMEI. Cysteine residues involved in disulfide bridges are indicated*

PMEI is a monomeric protein with a molecular weight of 16 kDa, as detected by SDS-PAGE. The complete amino acid sequence of PMEI was obtained by combining direct N-terminal sequencing of the entire protein, sequencing of peptides derived from acid cleavage of the two Asp-Pro bonds, and sets of peptides derived from trypsin, Asp-N and Lys-C endoproteinase digestions.[16] The total sequence comprises 152 amino acid residue (Figure 5), yielding a molecular weight of 16,277 Da, in good agreement with that obtained by SDS-PAGE. The sequence contains five Cys residues. Analysis of fragments obtained after digestion of the protein alkylated without previous reduction identifies two disulfide bridges connecting Cys9 with Cys18, and Cys 74 with Cys114, whereas Cys140 bears a free thiol group. The protein shows micro-heterogeneities at the N-terminus (part of the polypeptide chain has an additional Ala residue), and at five positions along the amino acid sequence (Ala/Ser56, Tyr/Phe78, Ser/Asn117, Asn/Asp123, and Val/Ile142). The presence of two forms, having or lacking the initial Ala residue, can be attributed to an incomplete processing of the protein, whereas the detection of different amino acid residues at well defined positions along the polypeptide chain can be attributed to the presence of multiple isoforms of the inhibitor in the kiwi fruit. HPLC analysis reveals the presence of at least 6 different isoforms (Figure 6 and 7).

The amino acid sequence alignment identifies homology with invertase inhibitors from *N. tabacum* and *L. esculentum*.[17] The overall amino acid residue identity is low, ranging between 25-30%, in particular four Cys residues are conserved. In PMEI these Cys residues are involved in two disulfide bridges, which play a considerable role in the

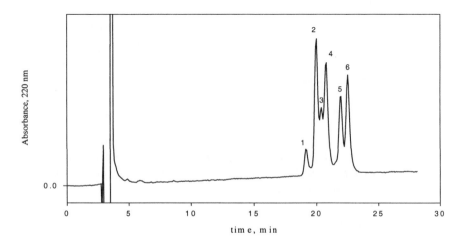

Figure 6 *HPLC analysis of PMEI isolated from Kiwi fruits*

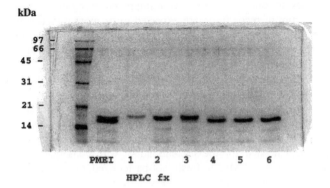

Figure 7

maintenance of the protein structure. It is possible to hypothesize that a common arrangement of disulfide bridges is shared by the homologous proteins, constituting a common three-dimensional fold.

The binding kinetics of the PME-PMEI interaction was studied by surface plasmon resonance analyzing various concentrations of PMEI on PME immobilized on a sensor chip. The interaction was measured in real time by recording the changes in Resonance Units (RU), which are proportional to the mass of protein binding. The equilibrium dissociation constant for the binding (K_D) was determined using two different buffer conditions: pH 7.0 or pH 6.0 (Figure 8). Interestingly the K_D measured at pH 7.0 (53 nM) had a value about ten fold higher than that observed at pH 6.0 (5.0 nM),. At lower pH, closer to that releaved in the apoplastic environment, the higher affinity of the interaction is mainly due to changes in the dissociation kinetics.

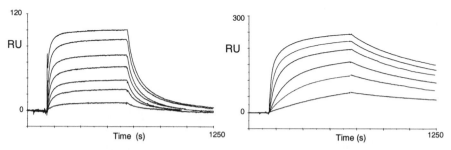

Figure 8 *Interaction between PMEI from Kiwi and immobilized PME from tomato at pH 7 (left) and pH 6 (right). Concentrations of PMEI used at pH 7, from bottom to top curve: 6 nM, 13 nM, 27 nM, 50 nM, 100 nM, 200 nM and 400 nM; pH6, from bottom to top curve: 3 nM, 10 nM, 20 nM, 50 nM, 100 nM*

4 CONCLUSIONS

Knowledge of the structural requirements that confer to inhibitors of pectic enzymes the ability of interacting specifically with their ligands can be exploited to devise in vitro mutagenesis-based strategies to obtain proteins with improved recognition abilities. PGIPs with a broader spectrum of recognition are expected to be more effective in the defense against pathogens. Improvement of resistance may be obtained by transforming plants either with a set of different *pgip* genes with complementary recognition abilities, or with single genes with multiple specificities.

The ability of PMEI to inhibit plant PMEs raises a potential interest of the inhibitor in the technology of fruit products. PME is responsible for many commercially deleterious effects, such as loss of consistence in solid products and diminution of viscosity and turbidity in fruit juices, ultimately resulting in the phase separation (cloud-loss) of the juice. In order to inactivate PME, many plant products are subjected to thermal treatments harsher than those needed for the pasteurization, with consequent loss of flavors and nutrients. The addition of PMEI to fresh, not-pasteurized, orange juices may preserve the cloud stability.[18] This makes PMEI very attractive as a technological adjuvant in the fruit juice stabilization.

References

1. De Lorenzo, G., D'Ovidio, R., and Cervone, F. (2001) *Annual Review of Phytopathology (in press)*.
2. Favaron, F., Castiglioni, C., and Di Lenna, P. (1993) *J.Phytopathol.* **139,** 201-206.
3. Favaron, F., Castiglioni, C., D'Ovidio, R., and Alghisi, P. (1997) *Physiol.Mol.Plant Pathol.* **50,** 403-417.
4. Bergmann, C., Ito, Y., Singer, D., Albersheim, P., Darvill, A. G., Benhamou, N., Nuss, L., Salvi, G., Cervone, F., and De Lorenzo, G. (1994) *Plant J.* **5,** 625-634.
5. Nuss, L., Mahé, A., Clark, A. J., Grisvard, J., Dron, M., Cervone, F., and De Lorenzo, G. (1996) *Physiol.Mol.Plant Pathol.* **48,** 83-89.
6. De Lorenzo, G., Cervone, F., Bellincampi, D., Caprari, C., Clark, A. J., Desiderio, A., Devoto, A., Forrest, R., Leckie, F., Nuss, L., and Salvi, G. (1994) *Biochem.Soc.Trans.* **22,** 396-399.
7. Cervone, F., De Lorenzo, G., Pressey, R., Darvill, A. G., and Albersheim, P. (1990) *Phytochem.* **29,** 447-449.
8. Desiderio, A., Aracri, B., Leckie, F., Mattei, B., Salvi, G., Tigelaar, H., Van Roekel, J. S. C., Baulcombe, D. C., Melchers, L. S., De Lorenzo, G., and Cervone, F. (1997) *Mol.Plant-Microbe Interact.* **10,** 852-860.
9. Kobe, B. and Deisenhofer, J. (1993) *Nature* **366,** 751-756.
10. Kobe, B. and Deisenhofer, J. (1995) *Nature* **374,** 183-186.
11. Leckie, F., Mattei, B., Capodicasa, C., Hemmings, A., Nuss, L., Aracri, B., De Lorenzo, G., and Cervone, F. (1999) *EMBO J.* **18,** 2352-2363.
12. Mattei, B., Bernalda, M. S., Federici, L., Roepstorff, P., Cervone, F., and Boffi, A. (2001) *Biochemistry* **40,** 569-576.
13. Caprari, C., Mattei, B., Basile, M. L., Salvi, G., Crescenzi, V., De Lorenzo, G., and Cervone, F. (1996) *Mol.Plant-Microbe Interact.* **9,** 617-624.
14. Balestrieri, C., Castaldo, D., Giovane, A., Quagliuolo, L., and Servillo, L. (1990) *Eur.J.Biochem.* **193,** 183-187.
15. Giovane, A., Balestrieri, C., Quagliuolo, L., Castaldo, D., and Servillo, L. (1995) *Eur.J.Biochem.* **233,** 926-929.
16. Camardella, L., Carratore, V., Ciardiello, M. A., Servillo, L., Balestrieri, C., and Giovane, A. (2000) *Eur.J.Biochem.* **267,** 4561-4565.
17. Greiner, S., Krausgrill, S., and Rausch, T. (1998) *Plant Physiol* **116,** 733-742.
18. Castaldo, D., Lovoi, A., Quagliuolo, L., Servillo, L., Balestrieri, C., and Giovane, A. (1991) *J.Food Sci.* **56,** 1632-1634.

7 Information Mining in Genomes and Glycomes

CARBOHYDRATE-ACTIVE ENZYMES IN COMPLETELY SEQUENCED GENOMES

Bernard Henrissat[1] and Pedro M. Coutinho[2]

[1]Architecture et Fonction des Macromolécules Végétales, UMR 6098, CNRS, Universités Aix-Marseille I and II, 31 Chemin Joseph Aiguier, F-13402 Marseille cedex 20, France
[2]Instituto Superior Tecnico, Centre for Biological & Chemical Engineering, Av. Rovisco Pais, P-1049-001 Lisboa, Portugal

SUMMARY

In 1991, we introduced a family classification system for glycoside hydrolases and transglycosidases based on amino acid sequence similarities. More recently we have introduced a similar classification system for glycosyltransferases. With the growing amount of sequence and biochemical data, the number of families of glycoside hydrolases and glycosyltransferases has grown over the years. There are presently over 80 families of glycoside hydrolases and more than 50 families of glycosyltransferases. These families can be accessed conveniently on a permanently updated www server at URL: http://afmb.cnrs-mrs.fr/~pedro/CAZY/db.html. Although it is likely that other families will be discovered, it is reasonable to assume that the most important families are already identified and that general inferences on the role and importance of carbohydrates in different organisms can be made. An advantage of the sequenced-based families is that they can be readily searched in genomes, allowing a global and detailed comparison of the glycoside hydrolase/glycosyltransferase repertoire of various organisms at a genomic scale. Here we have searched glycoside hydrolases and glycosyltransferases in 18 completely sequenced genomes covering Archaea, Bacteria and Eukaryotes.

1 INTRODUCTION

A unique aspect of oligo- and polysaccharides is their exceptional structural and functional diversity which is a challenge for the enzymes responsible for their selective cleavage (glycosidases; polysaccharide lyases), rearrangements (transglycosidases), biosynthesis (glycosyltransferases) and modification (for example, carbohydrate esterases). The multiplicity of these "carbohydrate-active enzymes" poses enormous problems for their classification. The IUB-MB EC numbers are based on the reaction catalysed and on the substrate specificity of the enzymes. This classification scheme neither (and was not intended to) reflects the structural features nor the catalytic mechanism of the enzymes. Furthermore, at the structural level, a widespread feature of carbohydrate-active enzymes is their modular organisation where the catalytic module often carries one or several ancillary modules (sometimes more than 10!) whose function can be carbohydrate-binding, catalytic, or simply unknown. The recent explosion of DNA sequences provided by genome-sequencing projects generates vast problems for the description and annotation of open-reading frames, since few if any of the ORFs will have received functional or structural analysis.

In order to rationalise the organisation of these diverse and complex enzymes, we introduced a new classification system, initially for the glycoside hydrolases (GHs),[1-3] and later for the glycosyltransferases (GTs),[4] based on the amino acid sequence similarities within the catalytic domains of these enzymes. In parallel, a similar system was introduced for the carbohydrate-binding modules.[5,6] This sequence classification has been rapidly and almost universally adopted by the community and forms the foundation for much of the chemical and structural "genomics" in the field (discussed further below). A number of crucial features of enzyme action, catalysis, evolution and 3-D structure are revealed by the sequence classification in a powerful predictive manner.

2 EVOLUTIONARY RELATIONSHIPS

One of the major initial advances provided by the family classification was the observation that there are a large number of 'polyspecific' enzyme families, i.e., families which contain enzymes of different and distinct substrate specificity. Conversely, it was fascinating to find that enzymes acting on the same substrate could belong to totally different families, frequently displaying unrelated 3-D folds.[1,7,8] These situations were quite surprising in view of the relatively small number of folds found for related systems such as lipases and proteinases. The sequence-family classification clearly indicates both divergent and convergent evolution and can direct and inform rational design strategies for the tailoring of enzyme specificity as shown for instance recently.[9,10]

3 ENZYME CATALYTIC MECHANISM

The catalytic mechanism of carbohydrate-active enzymes is dictated by their 3-D structure. Since 3-D structure is derived from sequence, the catalytic mechanism of an enzyme (including those derived from ORFs) is revealed by the sequence-family classification. For glycoside hydrolases, for example, the stereochemistry of hydrolysis (inversion or retention of anomeric configuration) is strictly conserved within a given family, as is the position of the participating groups in 3-D space. This observation has suffered no counterexample since it was first observed by Gebler and colleagues.[11] Furthermore, since enzyme inhibitors and transition-state mimics utilise prior knowledge of the catalytic mechanism, the sequence-family classification has underpinned the specific design and synthesis of therapeutic agents targeted at utilising structural features from related enzymes.[12,13]

4 THREE-DIMENSIONAL STRUCTURAL FEATURES

As expected from sequence-based families, it has been found that all the members of a given family share the same fold.[7,8] Because 3-D structures are better conserved than sequences, a higher level of hierachy (the 'clans'), between the families and the folds, was introduced to group families which displayed a similar fold together with an identical mechanism and catalytic machinery.[8,14] Since the 3-D structure of any protein sequence or ORF can now be predicted if a structural representive is known, the sequence-family classification has allowed the homology modelling of related enzymes within a given family. A fine example is the description of the molecular basis of Tay-Sachs disease based upon sequence similarities of the human hexosaminidase to a bacterial chiobiase.[15]

5 PREGENOMICS, GENOMICS AND POSTGENOMICS APPLICATIONS

The current classification has been rapidly and widely accepted by molecular biologists to assign sequences to established families, by microbiologists to design primers, by structural biologists to select targets for structural investigation and by enzymologists to rationalise mechanistic data.

With the advent of large scale genome sequencing projects, the sequence-based families of carbohydrate-active enzymes now allow efficient genome annotations, genome mining, and genome comparisons.[16,17] Here we examine the global content in glycoside hydrolase and glycosyltransferase of various genomes from free-living organisms from the three major kingdoms: archaea, bacteria and eukaryotes. For this, the potential glycoside hydrolases and glycosyltransferases of a genome can only be identified by sequence similarity with known enzymes using similarity search programs such as (PSI)-BLAST[18] or established family profiles.[19] The frequent modular structure of carbohydrate-active enzymes poses particular problems in such a task, because a significant similarity between a coding region and the sequence of a carbohydrate-active enzyme can be confined to a non-catalytic region and not with the catalytic domain itself. Such 'hits' with non-catalytic regions have led to numerous erroneous genome annotations for carbohydrate-active enzymes.[20] To avoid this problem we have compared each genome coding region to a library of > 16,000 fragments corresponding to isolated catalytic and non-catalytic modules of carbohydrate-active enzyme sequences.

The results of this search are summarised in Table 1. All the genomes examined contain potential glycosyltransferases, but a few archaeal genomes (*Aeropyrum pernix*, *Archaeoglobus fulgidus* and *Methanobacterium thermoautotrophicum*) appear to be totally devoid in identifiable glycoside hydrolases.[17] By contrast, the percentage of genes devoted to potential glycoside hydrolases and glycosyltransferases is remarkably narrow in bacteria and eukaryotes: ORFs with similarity to glycoside hydrolases and glycosyltransferases represent 1.4 – 1.5% of all coding regions for most genomes examined. Only *Arabidopsis* and *Thermotoga maritima* have significantly more (about 3%). Interestingly, *T. maritima* seems adapted to a vegetarian 'diet' with numerous glycoside hydrolases directed against the plant cell wall.

The examination of the precise pattern of family distribution across organisms or kingdoms – including examination of those families which are absent or underrepresented in the 18 genomes examined – will undoubtedly reveal the emerging secrets of higher carbohydrate metabolism during evolution.

Table 1 *Glycoside hydrolase and glycosyltransferase content of various completely sequenced genomes*

	Organism	Ref	Genes	GHs	GTs	Total	%
Archaea	Aeropyrum pernix	21	2694	0	5	5	0.2
	Archaeoglobus fulgidus	22	2407	0	16	15	0.6
	Methanobacterium thermoautotrophicum	23	1869	0	19	19	1.0
	Methanococcus jannaschii	24	1771	2	9	11	0.6
	Pyrococcus abyssi		1765	6	20	26	1.4
	Pyrococcus horikoshii	25	2064	7	15	22	1.0
Bacteria	Aquifex aeolicus	26	1522	4	19	23	1.5
	Bacillus subtilis	27	4100	40	30	70	1.7
	Deinococcus radiodurans	28	3123	14	17	31	1.0
	Escherichia coli K12	29	4289	33	28	61	1.4
	Haemophilus influenzae	30	1711	4	15	19	1.1
	Mycobacterium tuberculosis	31	3918	16	29	45	1.1
	Synechocystis sp.	32	3169	10	51	61	1.9
	Thermotoga maritima	33	1846	40	16	56	3.0
Eukaryotes	Arabidopsis thaliana	34	25498	379	356	735	2.8
	Caenorhabditis elegans	35	19099	78	201	279	1.4
	Drosophila melanogaster	36	14100	90	115	205	1.4
	Saccharomyces cerevisiae	37	6203	41	49	90	1.4

Ref: reference; **genes**: estimated number of genes in genome; **GHs**: number of potential glycoside hydrolases; **GTs**: number of potential glycosyltransferases; **total**: sum of GHs+GTs; **%**: percentage of genes devoted to glycoside hydrolases and glycosyltransferases.

Acknowledgements

The authors are particularly grateful to Gideon J. Davies for his help and insight throughout the years.

References

1. B. Henrissat, *Biochem. J.*, 1991, **280**, 309.
2. B. Henrissat and A. Bairoch, *Biochem. J.*, 1993, **293**, 781.
3. B. Henrissat and A. Bairoch, *Biochem. J.*, 1996, **316**, 695.
4. J. A. Campbell, G. J. Davies, V. Bulone and B. Henrissat, *Biochem J*, 1997, **326**, 929.
5. P. Tomme, R. A. J. Warren, R. C. J. Miller, D. G. Kilburn and N. R. Gilkes, in *Enzymatic Degradation of Insoluble Polysaccharides*, vol. 618, ed. J. N. Saddler and M. Penner, Amercian Chemical Society, Washington, 1995, p. 142-163.

6.   A. B. Boraston, B. W. McLean, J. M. Kormos, M. Alam, N. R. Gilkes, C. A. Haynes, P. Tomme, D. G. Kilburn and R. A. J. Warren, in *Recent Advances in Carbohydrate Bioengineering*, ed. H. J. Gilbert G. J. Davies B. Henrissat and B. Svensson, The Royal Society of Chemistry, Cambridge, 1999, p. 202-211.

7.   G. Davies and B. Henrissat, *Structure*, 1995, **3**, 853.

8.   B. Henrissat and G. Davies, *Curr. Opin. Struct. Biol.*, 1997, **7**, 637.

9.   V. Ducros, S. J. Charnock, U. Derewenda, Z. S. Derewenda, Z. Dauter, C. Dupont, F. Shareck, R. Morosoli, D. Kluepfel and G. J. Davies, *J. Biol. Chem.*, 2000, **275**, 23020.

10. S. R. Andrews, S. J. Charnock, J. H. Lakey, G. J. Davies, M. Claeyssens, W. Nerinckx, M. Underwood, M. L. Sinnott, R. A. Warren and H. J. Gilbert, *J. Biol. Chem.*, 2000, **275**, 23027.

11. J. Gebler, N. R. Gilkes, M. Claeyssens, D. B. Wilson, P. Béguin, W. W. Wakarchuk, D. G. Kilburn, R. C. Miller, Jr., R. A. Warren and S. G. Withers, *J. Biol. Chem.*, 1992, **267**, 12559.

12. T. D. Heightman and A. T. Vasella, *Ang. Chem. Int. Ed.*, 1999, **38**, 750.

13. A. Varrot, M. Schülein, M. Pipelier, A. Vasella and G. J. Davies, *J. Am. Chem. Soc.*, 1999, **121**, 2621.

14. B. Henrissat, I. Callebaut, S. Fabrega, P. Lehn, J. P. Mornon and G. Davies, *Proc. Natl. Acad. Sci. USA*, 1995, **92**, 7090.

15. I. Tews, A. Perrakis, A. Oppenheim, Z. Dauter, K. S. Wilson and C. E. Vorgias, *Nature Struct. Biol.*, 1996, **3**, 638.

16. P. M. Coutinho and B. Henrissat, in *Recent Advances in Carbohydrate Bioengineering*, ed. H. Gilbert G. Davies B. Henrissat and B. Svensson, The Royal Society of Chemistry, Cambridge, 1999, p. 3-12.

17. P. M. Coutinho and B. Henrissat, *J. Mol. Microbiol. Biotechnol.*, 1999, **1**, 307.

18. S. F. Altschul, T. L. Madden, A. A. Schaffer, J. Zhang, Z. Zhang, W. Miller and D. J. Lipman, *Nucleic Acids Res.*, 1997, **25**, 3389.

19. A. Bateman, E. Birney, R. Durbin, S. R. Eddy, K. L. Howe and E. L. Sonnhammer, *Nucleic Acids Res.*, 2000, **28**, 263.

20. B. Henrissat and G. J. Davies, *Plant Physiol.*, 2000, **124**, 1515.

21. Y. Kawarabayasi, Y. Hino, H. Horikawa, S. Yamazaki, Y. Haikawa, K. Jin-no, M. Takahashi, M. Sekine, S. Baba, A. Ankai, H. Kosugi, A. Hosoyama, S. Fukui, Y. Nagai, K. Nishijima, H. Nakazawa, M. Takamiya, S. Masuda, T. Funahashi, T. Tanaka, Y. Kudoh, J. Yamazaki, N. Kushida, A. Oguchi, H. Kikuchi et al., *DNA Res.*, 1999, **6**, 83.

22. H. P. Klenk, R. A. Clayton, J. F. Tomb, O. White, K. E. Nelson, K. A. Ketchum, R. J. Dodson, M. Gwinn, E. K. Hickey, J. D. Peterson, D. L. Richardson, A. R. Kerlavage, D. E. Graham, N. C. Kyrpides, R. D. Fleischmann, J. Quackenbush, N. H. Lee, G. G. Sutton, S. Gill, E. F. Kirkness, B. A. Dougherty, K. McKenney, M. D. Adams, B. Loftus, J. C. Venter et al., *Nature*, 1997, **390**, 364.

23. D. R. Smith, L. A. Doucette-Stamm, C. Deloughery, H. Lee, J. Dubois, T. Aldredge, R. Bashirzadeh, D. Blakely, R. Cook, K. Gilbert, D. Harrison, L. Hoang, P. Keagle, W. Lumm, B. Pothier, D. Qiu, R. Spadafora, R. Vicaire, Y. Wang, J. Wierzbowski, R. Gibson, N. Jiwani, A. Caruso, D. Bush, J. N. Reeve et al., *J. Bacteriol.*, 1997, **179**, 7135.

24. C. J. Bult, O. White, G. J. Olsen, L. Zhou, R. D. Fleischmann, G. G. Sutton, J. A. Blake, L. M. FitzGerald, R. A. Clayton, J. D. Gocayne, A. R. Kerlavage, B. A. Dougherty, J. F. Tomb, M. D. Adams, C. I. Reich, R. Overbeek, E. F. Kirkness, K. G. Weinstock, J. M. Merrick, A. Glodek, J. L. Scott, N. S. M. Geoghagen and J. C. Venter, *Science*, 1996, **273**, 1058.

25. Y. Kawarabayasi, M. Sawada, H. Horikawa, Y. Haikawa, Y. Hino, S. Yamamoto, M. Sekine, S. Baba, H. Kosugi, A. Hosoyama, Y. Nagai, M. Sakai, K. Ogura, R. Otsuka, H. Nakazawa, M. Takamiya, Y. Ohfuku, T. Funahashi, T. Tanaka, Y. Kudoh, J. Yamazaki, N. Kushida, A. Oguchi, K. Aoki and H. Kikuchi, *DNA Res*, 1998, **5**, 55.

26. G. Deckert, P. V. Warren, T. Gaasterland, W. G. Young, A. L. Lenox, D. E. Graham, R. Overbeek, M. A. Snead, M. Keller, M. Aujay, R. Huber, R. A. Feldman, J. M. Short, G. J. Olsen and R. V. Swanson, *Nature*, 1998, **392**, 353.

27. F. Kunst, N. Ogasawara, I. Moszer, A. M. Albertini, G. Alloni, V. Azevedo, M. G. Bertero, P. Bessieres, A. Bolotin, S. Borchert, R. Borriss, L. Boursier, A. Brans, M. Braun, S. C. Brignell, S. Bron, S. Brouillet, C. V. Bruschi, B. Caldwell, V. Capuano, N. M. Carter, S. K. Choi, J. J. Codani, I. F. Connerton, A. Danchin et al., *Nature*, 1997, **390**, 249.

28. O. White, J. A. Eisen, J. F. Heidelberg, E. K. Hickey, J. D. Peterson, R. J. Dodson, D. H. Haft, M. L. Gwinn, W. C. Nelson, D. L. Richardson, K. S. Moffat, H. Qin, L. Jiang, W. Pamphile, M. Crosby, M. Shen, J. J. Vamathevan, P. Lam, L. McDonald, T. Utterback, C. Zalewski, K. S. Makarova, L. Aravind, M. J. Daly, C. M. Fraser et al., *Science*, 1999, **286**, 1571.

29. F. R. Blattner, G. Plunkett, 3rd, C. A. Bloch, N. T. Perna, V. Burland, M. Riley, J. Collado-Vides, J. D. Glasner, C. K. Rode, G. F. Mayhew, J. Gregor, N. W. Davis, H. A. Kirkpatrick, M. A. Goeden, D. J. Rose, B. Mau and Y. Shao, *Science*, 1997, **277**, 1453.

30. R. D. Fleischmann, M. D. Adams, O. White, R. A. Clayton, E. F. Kirkness, A. R. Kerlavage, C. J. Bult, J. F. Tomb, B. A. Dougherty, J. M. Merrick et al., *Science*, 1995, **269**, 496.

31. S. T. Cole, R. Brosch, J. Parkhill, T. Garnier, C. Churcher, D. Harris, S. V. Gordon, K. Eiglmeier, S. Gas, C. E. Barry, 3rd, F. Tekaia, K. Badcock, D. Basham, D. Brown, T. Chillingworth, R. Connor, R. Davies, K. Devlin, T. Feltwell, S. Gentles, N. Hamlin, S. Holroyd, T. Hornsby, K. Jagels, B. G. Barrell et al., *Nature*, 1998, **393**, 537.

32. T. Kaneko, S. Sato, H. Kotani, A. Tanaka, E. Asamizu, Y. Nakamura, N. Miyajima, M. Hirosawa, M. Sugiura, S. Sasamoto, T. Kimura, T. Hosouchi, A. Matsuno, A. Muraki, N. Nakazaki, K. Naruo, S. Okumura, S. Shimpo, C. Takeuchi, T. Wada, A. Watanabe, M. Yamada, M. Yasuda and S. Tabata, *DNA Res.*, 1996, **3**, 109.

33. K. E. Nelson, R. A. Clayton, S. R. Gill, M. L. Gwinn, R. J. Dodson, D. H. Haft, E. K. Hickey, J. D. Peterson, W. C. Nelson, K. A. Ketchum, L. McDonald, T. R. Utterback, J. A. Malek, K. D. Linher, M. M. Garrett, A. M. Stewart, M. D. Cotton, M. S. Pratt, C. A. Phillips, D. Richardson, J. Heidelberg, G. G. Sutton, R. D. Fleischmann, J. A. Eisen, C. M. Fraser et al., *Nature*, 1999, **399**, 323.

34. T. A. G. Initiative, *Nature*, 2000, **408**, 796.

35. T. C. e. S. Consortium, *Science*, 1998, **282**, 2012.

36. M. D. Adams, S. E. Celniker, R. A. Holt, C. A. Evans, J. D. Gocayne, P. G. Amanatides, S. E. Scherer, P. W. Li, R. A. Hoskins, R. F. Galle, R. A. George, S. E. Lewis, S. Richards, M. Ashburner, S. N. Henderson, G. G. Sutton, J. R. Wortman, M. D. Yandell, Q. Zhang, L. X. Chen, R. C. Brandon, Y. H. Rogers, R. G. Blazej, M. Champe, B. D. Pfeiffer, K. H. Wan, C. Doyle, E. G. Baxter, G. Helt, C. R. Nelson, G. L. Gabor Miklos, J. F. Abril, A. Agbayani, H. J. An, C. Andrews-Pfannkoch, D. Baldwin, R. M. Ballew, A. Basu, J. Baxendale, L. Bayraktaroglu, E. M. Beasley, K. Y. Beeson, P. V. Benos, B. P. Berman, D. Bhandari, S. Bolshakov, D. Borkova, M. R. Botchan, J. Bouck, P. Brokstein, P. Brottier, K. C. Burtis, D. A. Busam, H. Butler, E. Cadieu, A. Center, I. Chandra, J. M. Cherry, S. Cawley, C. Dahlke, L. B. Davenport, P. Davies, B. de Pablos, A. Delcher, Z. Deng, A. D. Mays, I. Dew, S. M. Dietz, K. Dodson, L. E. Doup, M. Downes, S. Dugan-Rocha, B. C. Dunkov, P. Dunn, K. J. Durbin, C. C. Evangelista, C. Ferraz, S. Ferriera, W. Fleischmann, C. Fosler, A. E. Gabrielian, N. S. Garg, W. M. Gelbart, K. Glasser, A. Glodek, F. Gong, J. H. Gorrell, Z. Gu, P. Guan, M. Harris, N. L. Harris, D. Harvey, T. J. Heiman, J. R. Hernandez, J. Houck, D. Hostin, K. A. Houston, T. J. Howland, M. H. Wei, C. Ibegwam, M. Jalali, F. Kalush, G. H. Karpen, Z. Ke, J. A. Kennison, K. A. Ketchum, B. E. Kimmel, C. D. Kodira, C. Kraft, S. Kravitz, D. Kulp, Z. Lai, P. Lasko, Y. Lei, A. A. Levitsky ,J. Li, Z. Li, Y. Liang, X. Lin, X. Liu, B. Mattei, T. C. McIntosh, M. P. McLeod, D. McPherson, G. Merkulov, N. V. Milshina, C. Mobarry, J. Morris, A. Moshrefi, S. M. Mount, M. Moy, B. Murphy, L. Murphy, D. M. Muzny, D. L. Nelson, D. R. Nelson, K. A. Nelson, K. Nixon, D. R. Nusskern, J. M. Pacleb, M. Palazzolo, G. S. Pittman, S. Pan, J. Pollard, V. Puri, M. G. Reese, K. Reinert, K. Remington, R. D. Saunders, F. Scheeler, H. Shen, B. C. Shue, I. Siden-Kiamos, M. Simpson, M. P. Skupski, T. Smith, E. Spier, A. C. Spradling, M. Stapleton, R. Strong, E. Sun, R. Svirskas, C. Tector, R. Turner, E. Venter, A. H. Wang, X. Wang, Z. Y. Wang, D. A. Wassarman, G. M. Weinstock, J. Weissenbach, S. M. Williams, T. Woodage, K. C. Worley, D. Wu, S. Yang, Q. A. Yao, J. Ye, R. F. Yeh, J. S. Zaveri, M. Zhan, G. Zhang, Q. Zhao, L. Zheng, X. H. Zheng, F. N. Zhong, W. Zhong, X. Zhou, S. Zhu, X. Zhu, H. O. Smith, R. A. Gibbs, E. W. Myers, G. M. Rubin, and J. C. Venter, *Science*, 2000, **287**, 2185.

37. A. Goffeau, B. G. Barrell, H. Bussey, R. W. Davis, B. Dujon, H. Feldmann, F. Galibert, J. D. Hoheisel, C. Jacq, M. Johnston, E. J. Louis, H. W. Mewes, Y. Murakami, P. Philippsen, H. Tettelin and S. G. Oliver, *Science*, 1996, **274**, 563.

RECENT ADVANCES IN MYCOBACTERIAL ARABINOGALACTAN BIOSYNTHESIS IN THE POST-GENOMICS ERA

L. Kremer[1], L.G. Dover[2], S.S. Gurcha[2], A.K. Pathak[3], R.C. Reynolds[3] and G.S. Besra[2]

[1]INSERM U447, Institut Pasteur de Lille, 1 rue du Pr. Calmette, BP245-59019 Lille Cedex, France, [2]Department of Microbiology & Immunology, University of Newcastle, Newcastle upon Tyne NE2 4HH, [3]Department of Organic Chemistry, Southern Research Institute, P.O. Box 55305, Birmingham, AL 35255-5305, USA

1 GENERAL INTRODUCTION

Prevention efforts and control of tuberculosis is seriously hampered by the emergence of multi-drug resistant strains of *Mycobacterium tuberculosis*.[1,2] Therefore, new approaches to the treatment of tuberculosis are needed. Since the mycobacterial cell wall is essential for viability and virulence, it has become the focus for the search of essential targets in the development of new anti-mycobacterial agents.[3] The mycobacterial cell wall core consists of three interconnected macromolecules. Mycolic acids represent the outermost components of the cell wall. They are long chain α-alkyl-β-hydroxy fatty acids unique to mycobacteria and related taxa and are believed to play a crucial role in the architecture of the mycobacterial envelope.[4] The mycolic acids are further esterified to a heteropolysaccharide, arabinogalactan, which is composed primarily of D-arabinofuranosyl (Ara$f$) and D-galactofuranosyl (Gal$f$) residues. This entire polymer (mAG) is connected, *via* a unique linker disaccharide phosphate to a muramic acid residue of peptidoglycan (Figure 1). It is clear that, arabinogalactan plays a crucial role in the assembly of the mycobacterial cell wall and represents a suitable choice for the development of new targets leading to further drug development against pathogenic mycobacteria. This is supported by the fact that enzymes involved in the biogenesis and assembly of arabinogalactan are not found in humans. In addition, it has been established that several front-line drugs are known to target essential components of the mycobacterial cell wall. For instance, isoniazid (INH) is a potent inhibitor of mycolic acid biosynthesis targeting InhA[5,6] and possibly KasA.[7,8] Recently, pyrazinoic acid, the active form of the pro-drug pyrazinamide (PZA) was shown to target fatty acid synthase I from *M. tuberculosis*[9] which provides the short-chain fatty acid precursors for mycolic acid biosynthesis. Also, earlier studies demonstrated that ethambutol (EMB) led to a rapid inhibition of mycolic acid transfer to the cell wall and an accumulation of trehalose mono- and di-mycolates.[10,11] Subsequent studies have demonstrated that ethambutol disrupts the synthesis of the arabinan component of arabinogalactan by targeting various arabinosyltransferases, *embABC*.[12-15] Although the completion of the entire genome sequence of *M. tuberculosis*[16] greatly aids in the identification of enzymes involved in arabinogalactan biosynthesis, unambiguous identification from just sequence data is problematical. Therefore, genetic studies as well as biochemical approaches need to be developed in tandem. This communication focuses on recent advances in arabinogalactan biosynthesis in *M. tuberculosis* and also highlights the importance for the development of new tuberculosis therapeutics.

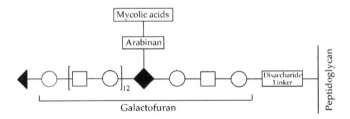

Figure 1 *Organization of the galactan region as well as its relationship to other major cell wall components* ◆ (5,6-β -D-Gal*f*), ○ (5-β -D-Gal*f*), □ (6-β -D- Gal*f*), ◀ (t-β - **D**- Gal*f*)

1.1 Structure of Arabinogalactan

As mentioned above the cell wall core is composed of a covalently linked complex of mycolic acids, **D**-arabinan and **D**-galactan attached to peptidoglycan *via* a α-L-Rha*p*-(1→3)-α-**D**-GlcNAc linkage unit (LU) and is often referred to as the mAGP complex. Detailed chemical characterization using gas-chromatography-mass spectrometry (GC-MS) and fast-atom bombardment mass spectrometry (FAB-MS)[17-20] revealed the structure was composed of (1) mycolic acids esterified at the 5-hydroxyl position of the penultimate and terminal Ara*f* residues; (2) two or three arabinan chains are attached to C-5 of some of the 6-linked β-**D**-Gal*f* residues; (3) the galactan region consisted of a linear alternating Gal*f* polymer of around 30 residues possessing both 5-linked β-**D**-Gal*f* and 6-linked β-**D**-Gal*f* glycosyl residues; (4) the galactan region of arabinogalactan was linked to the C-6 of some of the MurNGly residues of peptidoglycan (PG) *via* the LU. This structural arrangement offers some explanation to why arabinogalactan is essential for viability in *M. tuberculosis*, it basically holds together the outer mycolic acid-lipid layer to the peptidoglycan layer.

1.2 Biosynthesis of Arabinogalactan

Biochemical studies indicate that, initially LU synthesis involves the transfer of GlcNAc-1-P and Rha*p* from their respective sugar nucleotides (UDP-GlcNAc and dTDP-Rha) to form the polyprenol-P-P-GlcNAc (GL-1) and polyprenol-P-P-GlcNAc- Rha (GL-2) lipid intermediates.[21] These reactions are catalyzed, by the decaprenol-monophosphate-α-*N*-acetylglucosaminyltransferase (encoded by the *rfe* gene, *Rv1302*) and the rhamnosyltransferase (encoded by the *wbbL* gene, *Rv3265c*),[16] respectively.

Interestingly, O-antigen biosynthesis can be restored in an *E. coli wbbL* insertional mutant by Rv3265c, suggesting that this open reading frame does encode for a rhamnosyltransferase. Furthermore, a number of studies have defined the genetics and enzymology surrounding the synthesis of the nucleotide precursor dTDP-Rha involved in LU synthesis. Indeed, dTDP-Rha residues are synthesized by a single pathway requiring

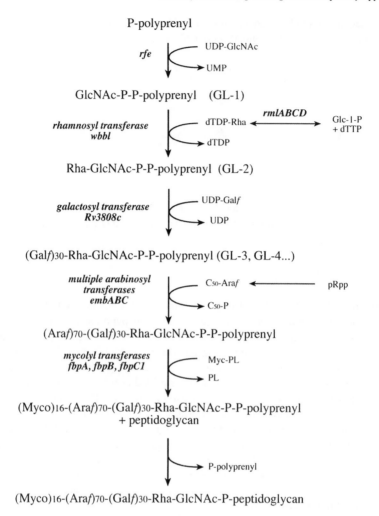

Figure 2 *Proposed pathway for cell wall biosynthesis in M. tuberculosis*

four enzymes (*rmlA* [*Rv0034*], *rmlB* [*Rv3464*], *rmlC* [*Rv3465*] and *rmlD* [*Rv3266c*]) beginning with TTP and Glc-1-P.[22-24] RmlA is a thymidylyltransferase, RmlB a dTDP-Glc dehydratase, RmlC a dTDP-6-deoxy-**D**-*xylo*-4-hexulose epimerase and RmlD a dTDP-6-deoxy-**L**-*lyxo*-4-hexulose reductase, respectively.[22-24]

The above glycolipid intermediates (GL-1 and GL-2) then serve as acceptors for the sequential addition of Gal*f* from UDP-Gal*f* to form polyprenol-P-P-GlcNAc-Rha-Gal$_{30}$.[21,25] Chemical analysis of the mature lipid-linked galactan synthesized *in vitro* suggests that this intermediate then serves as the acceptor for the subsequent addition of Ara*f* residues. β-D-Arabinofuranosyl-1-monophosphoryldecaprenol (DPA)[25-27] and phosphoribosyl pyrophosphate (PRPP)[28] are proposed to be the sugar donors implicated in the formation of the arabinan portion of arabinogalactan. None of the genes encoding for enzymes involved in DPA biosynthesis have been identified so far. This is probably reflective of the fact that Ara*f* is rarely found in nature and simple homology searching is proving inadequate in their subsequent identification. Addition of Ara*f* residues to the growing lipid intermediate, polyprenol-P-P-GlcNAc-Rha-Gal$_{30}$, ultimately leads to polyprenol-P-P-GlcNAc-Rha-Gal$_{30}$Ara$_{70}$, which at some point is mycolylated and transglycosylated to peptidoglycan (Figure 2). Based on the previous findings that the UDP-GlcNAc transferase is tunicamycin sensitive,[29] ethambutol[10-15] and isoniazid[5,6] target later steps involved in arabinan and mycolic acid biosynthesis, the intermediate steps in mAGP synthesis, notably galactan polymerization would represent novel drug targets.

2   GENETICS AND BIOCHEMISTRY OF GALACTAN BIOSYNTHESIS

2.1 Formation of the UDP-Galactofuranosyl Donor

UDP-Gal*f* consitutes the Gal*f* donor for galactan biosynthesis in mycobacteria. The early steps of the pathway involve the formation of UDP-Gal*p* from UDP-Glc by the UDP-Gal*p* epimerase (*galE* [Rv3634]). The enzyme, which forms UDP-Gal*f* from UDP-Gal*p*, the UDP-Gal*p* mutase has been extensively studied in *E. coli*[30] and *Klebsiella pneumoniae*.[31] More recent studies have led to the characterization of the *M. tuberculosis* UDP-Gal*p* mutase (*glf* [Rv3809]).[32]

2.2 *M. tuberculosis* Galactofuranosyltransferases

In order to develop an *in vitro* mycobacterial Gal*f* transferase assay, specific neoglycolipid acceptors, β-**D**-Gal*f*-(1→5)-β-**D**-Gal*f*-*O*-C$_8$ (Figure 3, G5G) and β-**D**-Gal*f*-(1→6)-β-**D**-Gal*f*-*O*-C$_8$ (Figure 3, G6G) were synthesized, respectively corresponding to the two major structural motifs found within the galactan of AG.[20]

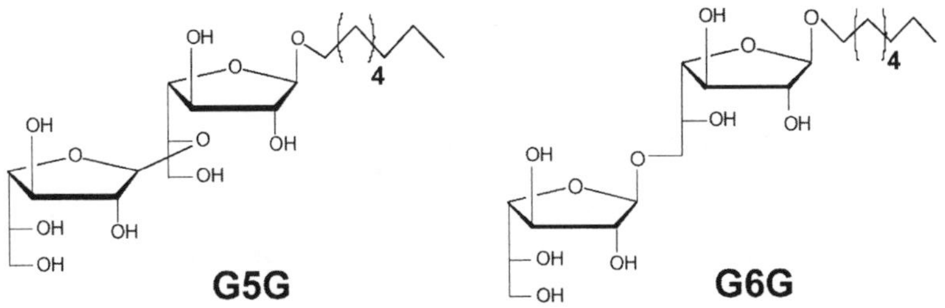

Figure 3 *Neoglycolipid acceptors*

TLC/autoradiography clearly demonstrated the enzymatic conversion of the both disaccharide acceptors to their corresponding trisaccharide (and tetrasaccharide) products in the presence of mycobacterial membranes, UDP-[^{14}C]Gal*p*, following endogenous conversion to UDP-[^{14}C]Gal*f* and subsequent galactosyltransferase activity (Figure 4). Additional assays performed in the presence of mycobacterial membranes and the cell wall enzymatic fraction P60 resulted in significant [^{14}C]Gal*f* incorporation from

UDP-[^{14}C]Gal*p*, following endogenous conversion to UDP-[^{14}C]Gal*f* and transferase activity for both β-**D**-Gal*f*-(1→5)-β-**D**-Gal*f*-*O*-C$_8$ and β-**D**-Gal*f*-(1→6)-β-**D**-Gal*f*-*O*-C$_8$ acceptors in a dose response (Figure 5). Interestingly, assays performed with P60 alone resulted in very poor [^{14}C]Gal*f* incorporation using both acceptors, however, provided a synergistic effect when added with membranes as compared to membranes alone. This may be correlated with the higher specific activity being observed for UDP-Gal*p* (*glf*) mutase activity within P60 preparations, resulting in a greater pool of UDP-Gal*f* for the subsequent galactosyltransferase(s).

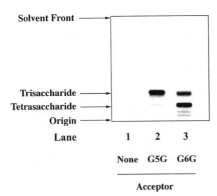

Figure 4 *Autoradiogram of reaction products produced through inclusion of acceptors G5G and G6G at 4 mM with M. smegmatis membranes and UDP-[^{14}C]Gal*

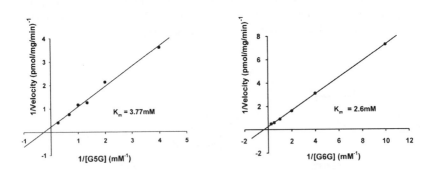

Figure 5 *Kinetic analysis of acceptors G5G and G6G*

2.3 Identification of a Bi-Functional Galactofuranosyltransferase

Since the Gal*f* transferases form rather different end products in *M. tuberculosis* than those found in *E. coli* and *Klebsiella*, it was very unlikely to find the *M. tuberculosis* Gal*f* transferase(s) by a simple homology search. The sensitive hydrophobic cluster analysis (HCA) method has been successfully used in several cases for the grouping of proteins of very low sequence similarity. Therefore, by mining the *M. tuberculosis* database using a combination of BLAST and HCA analyses, we identified several putative β-glycosyltransferases candidates (*Rv0539*, *Rv1208*, *Rv1500*, *Rv3631*, *Rv1518*, *Rv1541*, *Rv2957*, *Rv3631*, *Rv3782*, *Rv3808c*). Interestingly, one of them, *Rv3808c* (Figure 6), was located immediately downstream from the UDP-Gal*p* mutase *glf* (*Rv3809c*). The first four nucleotides of *Rv3808c* and the last four of *Rv3809c* overlapped and suggested that *Rv3808c* was possibly a β-galactosyltransferase transcriptionally coupled to the *glf* gene.[33]

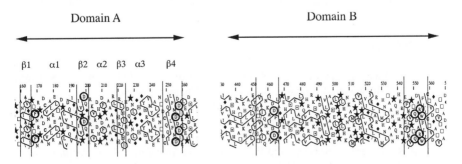

Figure 6 *Hydrophobic cluster analysis (HCA) of Rv3808c illustrating key domains. Strictly conserved residues are highlighted by bold circles*

Overexpression of *Rv3808c* in both *E. coli* and *M. smegmatis*, resulted in enhanced galactosyltransferase activity, by transferring [^{14}C]Gal from UDP-[^{14}C]Gal*f* to both β-**D**-Gal*f*-(1→5)-β-**D**-Gal*f*-*O*-C$_{10:1}$ and β-**D**-Gal*f*-(1→6)-β-**D**-Gal*f*-*O*-C$_{10:1}$ acceptors. Chemical analysis of the reaction products demonstrated that the β-**D**-Gal*f*-(1→5)-β-**D**-Gal*f*-*O*-C$_{10:1}$ acceptor resulted in a trisaccharide product β-**D**-Gal*f*-(1→6)-β-**D**-Gal*f*-(1→5)-β-**D**-Gal*f*-*O*-C$_{10:1}$, whereas the β-**D**-Gal*f*-(1→6)-β-**D**-Gal*f*-*O*-C$_{10:1}$ acceptor gave rise to both a trisaccharide β-**D**-Gal*f*-(1→5)-β-**D**-Gal*f*-(1→6)-β-**D**-Gal*f*-*O*-C$_{10:1}$ and a tetrasaccharide β-**D**-Gal*f*-(1→6)-β-**D**-Gal*f*-(1→5)-β-**D**-Gal*f*-(1→6)-β-**D**-Gal*f*-*O*-C$_{10:1}$, consistent with the expected alternating linkage profile of the galactan region of arabinogalactan.[33,34]. Interestingly, *Rv3808c* was found to be present in a "cell wall biosynthetic cluster".[16] Other open reading frames of interest in this region include the *emb* cluster encoding arabinofuranosyltransferases,[13,15] as well as, the 85A and 85C antigens that have been shown to exhibit mycolyltransferase activity.[35] In addition, interspersed throughout this region are a number of open reading frames encoding proteins with similarities to other polysaccharide based glycosyltransferase enzymes and ABC transporters, presumably involved in the assembly and final location of the arabinogalactan polymer in the cell wall.

In summary, we have demonstrated that, in association with the neoglycolipid assay, *Rv3808c*, now designated *glfT*, encodes a novel bifunctional enzyme, which performs two glycosyltransferases, UDP-Gal*f*:β-**D**-(1→5)galactofuranosyltransferase and UDP-Gal*f*:β-**D**-(1→6)galactofuranosyltransferase involved in cell wall galactan polymerization. Thus, GlfT represents a new cell wall drug target to be exploited in drug discovery programs leading to novel chemotherapeutics targeting *M. tuberculosis*. However, further experiments are required to investigate whether one or two distinct catalytic sites are responsible for these two galactofuranosyltransferase activities. In addition, a second more specialized galactofuranosyltransferase may participate in the transfer of Gal*f* to the Rha*p* residue of the LU.

Acknowledgments

GSB, Lister Institute-Jenner Research Fellow acknowledges support from GlaxoSmithKline Research & Development (UK) ActionTB Programme, The Medical Research Council (49343 and 49342), The Wellcome Trust (058972) and the National Institutes of Health USA (AI-45317)

References

1   D.A. Enarson and J.F. Murray, in *Tuberculosis*, eds. W.N. Rom and S. Garay, Little, Brown and Company, Boston, 1996, Chapter 4, p. 57.

2   L. Kremer and G.S. Besra, in *Recent Research Developments in Antimicrobial Agents & Chemotherapy*, ed. S.G. Pandalai, Research Signpost, Trivandrum-India, 1999, p. 453.

3   A.R. Baulard, G.S. Besra and P.J. Brennan, in *Mycobacteria: molecular biology and virulence*, eds. C. Ratledge and J. Dale, Blackwell Science, Oxford, 1999, Chapter 13, p. 240.

4   P.J. Brennan and H. Nikaido, *Annu. Rev. Biochem.*, 1995, **64**, 29.

5   A. Banerjee, E. Dubnau, A. Quemard, V. Balasubramanian, T. Um, K.S. Wilson, D. Collins, G. de Lisle and W. R. Jacobs, Jr., *Science*, 1994, **263**, 227.

6   L. Kremer, A.R. Baulard and G.S. Besra, in *Molecular Genetics of Mycobacteria*, eds. G. F. Hatfull and W. R. Jacobs, ASM Press, Washington D. C., 2000, Chapter 11, p. 173.

7   K. Mdluli, R.A. Slayden, Y.Q. Zhu, S. Ramaswamy, X. Pan, D. Mead, D.D. Crane, J.M. Musser and C.E. Barry III, *Science*, 1998, **280**, 1607.

8   R.A. Slayden, R.E. Lee and C.E. Barry III., *Mol. Microbiol.*, 2000, **38**, 514.

9   O. Zimhony, J.S. Cox, J.T. Welch, C. Vilcheze and W.R. Jacobs Jr., *Nat. Med.*, 2000, **6**, 1043.

10   K. Takayama, E.L. Armstrong, K.A. Kunugi and J.O. Kilburn, *Antimicrob. Agents Chemother.*, 1979, **16**, 240.

11   K. Takayama and J.O. Kilburn, *Antimicrob. Agents Chemother.*, 1989, **33**, 1493.

12   K. Mikusova, R.A. Salyden, G.S. Besra and P.J. Brennan, *Antimicrob. Agents Chemother.*, 1995, **39**, 2484.

13   A.E. Belanger, G.S. Besra, M.E. Ford, K. Mikusova, J.T. Belisle, P.J. Brennan and J.M. Inamine, *Proc. Natl. Acad. Sci. U. S. A.*, 1996, **93**, 11919.

14   L. Deng, K. Mikusova, K.G. Robuck, M.S. Sherman, P.J. Brennan and M.R. McNeil, *Antimicrob. Agents Chemother.*, 1995, **39**, 694.

15   A. Telenti, W.J. Philipp, S. Sreevatsan, C. Bernasconi, K.E. Stockbauer, B. Wieles, J.M. Musser and W.R. Jacobs, Jr., *Nat. Med.*, 1997, **3**, 567.

16 S.T. Cole, R. Brosch, J. Parkhill, T. Garnier, C. Churcher, D. Harris, S. V. Gordon, K. Eiglmeier, S. Gas, C. E. Barry III, F. Tekaia, K. Badcock, D. Basham, D. Brown, T. Chillingworth, R. Connor, R. Davies, K. Devlin, T. Feltwell, S. Gentles, N. Hamlin, S. Holroyd, T. Hornsby, K. Jagels, A. Krogh, J. McLean, S. Moule, L. Murphy, K. Oliver, J. Osborne, M. A. Quail, M-A. Rajandream, J. Rogers, S. Rutter, K. Seeger, J. Skelton, S. Squares, R. Squares, J. E. Sulston, K. Taylor, S. Whitehead and B. G. Barrell, *Nature*, 1998, **393**, 537.

17 M. Daffé, P. J. Brennan and M.R. McNeil, *J. Biol. Chem*, 1990, **265**, 6734.

18 M.R. McNeil, M. Daffé and P.J. Brennan, *J. Biol. Chem.*, 1990, **265**, 18200.

19 M.R. McNeil, M. Daffé and P. J. Brennan, *J. Biol. Chem.,* 1991, **266**, 13217.

20 G.S. Besra, K.H. Khoo, M.R. McNeil, A. Dell, H.R. Morris and P.J. Brennan, *Biochemistry*, 1995, **34**, 4257.

21 K. Mikusova, M. Mikus, G.S. Besra, I.C. Hancock and P.J. Brennan, *J. Biol. Chem.*, 1996, **271**, 7820.

22 Y. Ma, J.A. Mills, J.T. Belisle, V. Vissa, M. Howell, K. Bowlin, M.S. Scherman and M.R. McNeil, *Microbiology*, 1997, 143, 937.

23 R.J. Stern, T.Y. Lee, T.J. Lee, W. Yan, M.S. Scherman, V.D. Vissa, S.K. Kim, B.L. Wanner and M.R. McNeil, *Microbiology*, 1999, **145**, 663.

24 Y. Ma, R.J. Stern, M.S. Scherman, V.D. Vissa, W. Yan, V.C. Jones, F. Zhang, S.G. Franzblau, W.H. Lewis and M.R. McNeil, *Antimicrob. Agents Chemother.*, 2001, **45**, 1407.

25 G.S. Besra and P.J. Brennan, *Biochem. Soc. Trans.*, 1997, **25**, 845.

26 B.A. Wolucka, M.R. McNeil, E. de Hoffmann, T. Chojnacki and P.J. Brennan, *J. Biol. Chem.*, 1994, **269**, 23328.

27 R.E. Lee, K. Mikusova, P.J. Brennan and G.S. Besra, *J. Am. Chem. Soc.*, 1995, **117**, 11829.

28 M.S. Scherman, L. Kalbe-Bournonville, D. Bush, Y. Xin, L. Deng and M.R. McNeil, *J. Biol. Chem.*, 1996, **271**, 29652.

29 J.B. Ward and C.A. Curtis, *Eur. J. Biochem.*, 1992, **122**, 125.

30 P.M. Nassau, S.L. Martin, R.E. Brown, A. Weston, D. Monsey, M.R. McNeil and K. Duncan, *J. Bacteriol.*, 1996, **178**, 1047.

31 R. Koplin, J.R. Brisson and C. Whitfield, *J. Biol. Chem.*, 1997, **272**, 4121.

32 A. Weston, R.J. Stern, R.E. Lee, P.M. Nassau, D. Monsey, S.L. Martin, M.S. Scherman, G.S. Besra, K. Duncan and M.R. McNeil, *Tuberc. Lung Dis.*, 1998, **78**: 123.

33 K. Mikusova, T. Yagi, R. Stern, M.R. McNeil, G.S. Besra, D.C. Crick and P.J. Brennan, *J. Biol. Chem.*, 2000, **275**, 33890.

34 L. Kremer, L.G. Dover, C. Morehouse, P. Hitchin, H.R. Morris, A. Dell, P.J. Brennan, M.R. McNeil, C. Flaherty and G.S. Besra, *J. Biol. Chem.,* 2001, in press.

35 J.T. Belisle, V.D. Vissa, T. Sievert, K. Takayama, P.J. Brennan and G.S. Besra, *Science*, 1997, **276**, 1420.

NEOGLYCOLIPID TECHNOLOGY – AN APPROACH TO DECIPHERING THE INFORMATION CONTENT OF THE GLYCOME

T. Feizi

The Glycosciences Laboratory
Faculty of Medicine
Imperial College of Science, Technology and Medicine
Northwick Park Institute for Medical Research
Harrow HA1 3UJ

1 ABSTRACT

The neoglycolipid (NGL) technology was designed as a means of probing directly the roles of oligosaccharide sequences as antigens and ligands, and hence as a means of deciphering the information in the glycome. I cite here some examples of ways in which the technology has enabled assignments to be made of roles for oligosaccharide sequences as recognition elements for carbohydrate-recognizing receptors and antibodies, and has also led to the detection and characterization of hitherto unsuspected oligosaccharide sequences. The NGL principle is *par excellence* adaptable for modern micro-array technologies to fathom the prevalence and specificities of oligosaccharide-recognizing proteins in the proteome.

2 INTRODUCTION

In an article that I wrote for the Millennium Issue of the Glycoconjugate Journal,[1] I highlighted how the closing years of the second millennium had been truly uplifting for the field of carbohydrate biology, glycobiology. Our optimism that oligosaccharide sequences of glycoproteins and glycolipids are bearers of crucial biological information[2,3] been well borne out. This has been through a constellation of efforts of carbohydrate chemists, immunochemists, and cell- and molecular biologists. Thus, it is now unquestioned that specific oligosaccharide sequences are directly involved in protein targeting and folding, certain mechanisms of infection (host-parasite interactions), inflammation and immunity (both innate and adaptive).

In the mean time, it has emerged that the number of proteins encoded in the human genome is much less than anticipated: only in the order of 30 to 50 thousand. This has served to emphasize the importance of oligosaccharide chains as modulators of the activities and functions of the proteins in health and disease, through recognition processes mediated by carbohydrate-recognizing receptors. After the *genome* and the *proteome*, there has been the coining of the term *glycome* for the repertoire of oligosaccharide structures in the organism.

At a recent Gordon conference I was asked to hazard an estimate of the size of the glycome. I was taken aback as I had not really thought about an estimate; it must be enormous and almost unfathomable, as some oligosaccharide chains encompass dozens of monosaccharides, with diverse linkages, sequences, anomeric, configurations in their core, backbone and peripheral regions. There also occur additional substituents; and there are dramatic changes during embryonic development and cell differentiation. I question the usefulness of making a precise estimate of the size of the glycome. In my opinion, the challenge is to determine the specific roles, if any, of individual oligosaccharides or of particular domains on them, and to assign sequence-specificity. Indeed with the emergence of families of receptors with carbohydrate-binding activities (lectins), assignments of information content for defined oligosaccharide sequences will become more common, although the pinpointing and elucidation of bioactive domains on oligosaccharides is likely to remain one of the most challenging areas of cell biology. Ligands that are oligosaccharides cannot be readily cloned, as each is the product of multiple glycosyltransferases.

Here I discuss the neoglycolipid (NGL) technology, established with my colleagues,[4-6] which incorporates some of the key requirements for this challenge in the post-genomic era; and I highlight some examples of ways in which we have applied the technology to the oligosaccharide chains of glycoproteins, glycolipids, proteoglycans, whole cells and organs, in the search for ligands for carbohydrate-recognizing receptors of the immune system, and also in the definition of carbohydrate antigenic markers of biological interest.

3 PRINCIPLES OF THE NGL TECHNOLOGY

The technology involves the release of oligosaccharides (or their fragments) from the carrier molecules, and their conversion into lipid-linked oligosaccharide probes. The technology is also eminently applicable to desired chemically synthesized oligosaccharides. By a micro-derivatization procedure, the oligosaccharides are linked to an amino-phospholipid by reductive amination.[4,5] Complex mixtures of the lipid-linked oligosaccharide probes can be resolved and immobilized on matrices for ligand-binding experiments, in conjunction with sequence determination by mass spectrometry.[6,7] With the recent introduction[8] of a fluorochrome, anthracene, into the lipid tag as in structure **1**, the complex mixtures can be subjected to liquid chromatographies, with sensitive detection of the separation profiles. Given a well-expressed and folded carbohydrate-binding protein or a carbohydrate-recognizing antibody, the technology has the potential to find the ligand, even if this is a 'needle in the haystack' in a hugely heterogeneous glycan library. An important feature of the technology is that structural information can be derived on bioactive oligosaccharides without the need for preconceived ideas, the mass spectrometric readout being of key importance for structural assignment.

N-aminoacetyl-*N*-(9-anthracenylmethyl)-1,2-dihexadecyl-*sn*-glycero-3-
phosphoethanolamine

1

4 SOME APPLICATIONS AND FINDINGS

4.1 Novel Sulphated Ligands for the Selectins Revealed with NGLs from an Epithelial Mucin O-glycan Library

During 1990-91, following the finding that the leukocyte-endothelium adhesion molecules, E- and P-selectins, contain modules of lectin type, and with the knowledge that they bind to leukocytes of myeloid type, attention became focused on oligosaccharide sequences of the Lewis^x (Le^x) and Le^a series.[9] This was prompted by the knowledge that oligosaccharides of the Le^x series are differentiation antigens of myeloid cells.[10] By various approaches in different laboratories, including inhibition experiments with Le^x-related glycoconjugates and transfections of cells with glycosyltransferase genes, it was readily established sialyl oligosaccharides of the Le^x and Le^a series are ligands for the selectins.[11,12] The Le^a analogues are candidate ligands on human epithelial cells, as blood cells in the human do not produce Le^a oligosaccharides. We had an opportunity to examine a mucin type glycoprotein isolated from an ovarian cystadenoma for its content of E-selectin ligands.[13]. In the laboratory of Walter Morgan and Winifred Watkins, this glycoprotein had been noted to be unusually rich in sialic acid and the possibility was raised that it might support the binding of E-selectin This was readily corroborated by binding experiments with the glycoprotein immobilized in microwells. When we generated NGLs from the *O*-glycans of the glycoprotein, and probed these for E-selectin binding, large numbers of ligand-bearing oligosaccharide chains were indeed revealed. By mass spectrometry, however, these were found to be a family of sulphated rather than sialylated Le^x and Le^a chains such as oligosaccharides **2-5**. (Refs [13,14])

Thus, it was established that, for E-selectin binding, sulphate can substitute for the carboxyl group of sialic acid substituent on the Le^x and Le^a sequences.[13,15] The L- and P-selectins were later shown to also accommodate the sulpho-Le^x and -Le^a sequences.

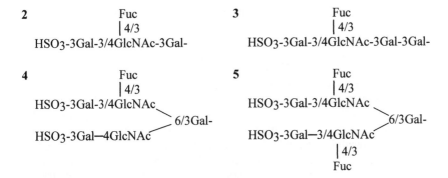

4.2 Unique Tetrasaccharide Sequence Revealed in a NGL Library from Heparan Sulphate

NGL technology is also eminently applicable for exploring bioactive oligosaccharide domains on polysaccharides. We are currently investigating the recognition elements of a monoclonal antibody, 10E4, which recognizes a heparan sulphate-related antigenic determinant that is closely associated with the brain lesions in prion disease.[16] For this purpose, we have generated, by limited digestion with heparan lyase III, an array of oligosaccharide fragments from an antigen-positive preparation of heparan sulphate. NGLs have been prepared from the fragments after their fractionation according to size, and these have been probed with 10E4 antibody. Among multiple antigen-positive components detected, a tetrasaccharide sequence has been partially characterized by mass spectrometry, and found to be the unique, non-sulphated sequence **6**, containing an *N*-unsubstituted glucosamine.[16] Clearly the outer disaccharide sequence is immuno-dominant as shown by the results of antigenic analyses of NGLs derived from the oligosaccharides **6-12**:

6	ΔUA-GlcN-UA-GlcNAc	positive
7	ΔUA-GlcN	positive
8	GlcN-UA-GlcNAc	negative
9	ΔUA-GlcNAc-UA-GlcNAc	negative
10	ΔUA-GlcNS-UA-GlcNAc	negative
11	GlcNAc-UA-GlcNAc	negative
12	GlcNS-UA-GlcNAc	negative

The 10E4 antibody clearly requires a terminal uronic acid, and can bind to the unsaturated uronic acid (ΔUA) generated in the course of the action of the lyase. The biosynthetic mechanisms and the possible involvement of the 10E4-active tetrasaccharide sequence in the course of prion lesions now require investigation.

4.3 Detection of Selectin Ligands with NGL Probes from Whole Cells

NGL technology is applicable to oligosaccharide populations derived from whole cells, as was illustrated with a cultured cell line. As in the human, the murine myeloid cells clearly bind to E-selectin, but they differ from those of the human in that the Le^x- and sialyl-Le^x-related antigens cannot be readily detected on them with the antibodies available. We examined a myeloid cell line of murine origin for the presence of ligand-positive oligosaccharide sequences.[17] The approach was to generate NGLs from oligosaccharides released from the surface of these cells with endo-β-galactosidase, and to probe these for E-selectin binding after resolving on thin layer chromatograms. Multiple ligand-bearing components were detected, and shown by mass spectrometry to be fuco-oligosaccharides of sialyl-Le^a/-Le^x type. Immunoreactivity with a monoclonal anti-sialyl-Le^a, showed that, in contrast to human myeloid cells, the rodent cells contain Le^a-related sequences.[17] It is now well established that Le^x-related sequences constitute major selectin ligands in murine blood cells.[18] The reason for our not detecting the sialyl-Le^x analogues among the oligosaccharide probes is that the anti-sialyl-Le^x, CSLEX1, used as reagent has a requirement for oligosaccharides with longer backbones than those released by the endo-β-galactosidase (T. Feizi and colleagues, unpublished observation).

4.4 A Novel Selectin Ligand and a Novel Class of Alkali-labile Oligosaccharides Revealed with NGL Libraries from Whole Organs

Two applications of NGL technology to glycans from whole organs further highlight the potential of this approach for exploring the information content of the glycome. In the first, glycolipids from murine kidney tissue were the source of oligosaccharides.[17] Having observed an unexpectedly strong binding of E-selectin to a heterogeneous population of neutral glycolipids extracted from the kidneys of mice (BALB/c strain), we released the oligosaccharides using endo-glycoceramidase, and converted these to NGLs. In contrast to the polydispersity of the starting glycolipid fraction, the derived NGLs migrated on TLC as a single band. This indicated that the polydispersity of the glycolipid fraction is a reflection of the heterogeneity of the lipid moiety. The NGLs could be analyzed by mass spectrometry more sensitively than the natural glycolipids, and the non-acidic ligand-positive oligosaccharide was thus identified as the Le^x-active globo-series sequence, **13**:

13 Galβ1-4GlcNAcβ1-6GalNAcβ1-3Galα1-4Galβ1-4Glc

| |

Fucα1.3.   Galβ1,3

The second organ-derived glycan library investigated as NGLs was from total brain glycoproteins.[19] Here the total glycans released by reductive alkaline borohydride degradation were investigated for the expression of the developmentally regulated carbohydrate antigen HNK-1. Not only were large arrays of antigen-positive glycans thus identified, but also, they were found to terminate with mannitol rather than *N*-acetylgalactosaminitol. The component **14** investigated in greatest detail by mass spectrometry was found to be based on a lactosamine backbone and to terminate in a 2-substituted mannitol:

    14 HSO_3-3GlcA1-3Gal1-4GlcNAc1-2Manol

A close examination of antigen-negative oligosaccharide alditols in the brain glycan library revealed that a substantial proportion (about 30%) were mannitol terminating, with 2- or 2,6-substituted mannitol[20] as in oligosaccharides **15-20**.

15	GlcNAcβ-2Manol
16	Galβ-4GlcNAcβ-2Manol
	Galβ-4GlcNAcβ-2Manol
17	|3
	Fucα
18	NeuAcα-3Galβ-4GlcNAcβ-2Manol

    ±NeuAcα-3Galβ-4GlcNAcβ-6⟍

19   Manol

    ±NeuAcα-3Galβ-4GlcNAcβ-2⟋

Mannose-terminating glycans from this source had been detected earlier, but thought to be 3-substituted.[21] The unequivocal assignment of 2-substitution by the NGL technology indicates that *O*-glycosylation of the type previously thought to occur in yeasts also occurs in the brain in higher eukaryotes.

4.5 Ligand Assignments Using Structurally Defined Oligosaccharides

Collections of NGLs from families of structurally defined oligosaccharides have been valuable in determining details of the sequence specifics of the selectins[22-24], and of carbohydrate-recognizing antibodies.[25] A recent application has been in identifying novel ligands for the cysteine-rich domain of the macrophage endocytosis receptor.[26] This carbohydrate-binding domain had earlier been shown to bind to the *N*-glycans of the pituitary hormone lutropin that terminate in 4-sulphated *N*-acetylgalactosamine.[27] Using

20	GalNAcβ1-4GlcNAcβ1- |4 HSO_3	Lutropin
21	GalNAcβ1-4GlcA/IdoAβ1- |4 HSO_3	Chondroitin 4-sulphates
22	Galβ1-3/4GlcNAcβ1- |3 HSO_3	Sulphated blood group chains

NGLs derived from a series of structurally defined oligosaccharides, two sets of additional biosynthetically distinct sequences were found to be bound, oligosaccharides **21** and **22**; these are chondroitin 4-sulphates and 3-sulphated blood group chains.[26] The structural basis of these cross-reactivities has been elucidated by x-ray crystallography of the protein module in complex with sulphated ligands.[28] The macrophage receptor is an antigen-presenting molecule Our findings raise interesting questions about the involvement of the cysteine-rich domain in immunity and autoimmunity.

5 PERSPECTIVES

The examples cited here of NGL probes from glycoproteins, glycolipids, proteoglycans, cells and organs are by no means an exhaustive survey. Some applications to N-glycans have been discussed elsewhere.[29-33] Current applications of the 'neo-lipid' principle include a peptide-based ligand for P-selectin, sulpho-tyrosine, which is bound in synergy with the carbohydrate ligand sialyl-Le^x, such that potent inhibitory analogues are generated when the two ligand classes linked to lipid are presented on liposomes.[34] Additional chemistries can be envisaged for generating NGLs from natural glycoconjugates in order to avoid ring-opening of the monosaccharide at the lipid-linkage, and leaving intact any recognition elements in the oligosaccharide core region. In the future, micro-array designs are envisaged based on NGLs, and used in conjunction with advanced protein expression systems, for the mapping of oligosaccharide-recognizing proteins in the proteome.

6 REFERENCES

1. T. Feizi, *Glycoconj.J*, 2000, **17**, 553.
2. T. Feizi, *Blood Trans.Immunohaematol.*, 1980, **23**, 563.
3. T. Feizi, *Nature*, 1985, **314**, 53.
4. P. W. Tang, H. C. Gooi, M. Hardy, Y. C. Lee and T. Feizi, *Biochemica Biophysica Research Communications*, 1985, **132**, 474.
5. M. S. Stoll, T. Mizuqchi, R. A. Childs and T. Feizi, *Biochemical Journal*, 1988, **256**, 661.
6. T. Feizi, M. S. Stoll, C.-T. Yuen, W. Chai and A. M. Lawson, *Methods Enzymol.*, 1994, **230**, 484.
7. A. M. Lawson, W. Chai, G. C. Cashmore, M. S. Stoll, E. F. Hounsell and T. Feizi, *Carbohydr.Res.*, 1990, **200**, 47.
8. M. S. Stoll, T. Feizi, R. W. Loveless, W. Chai, A. M. Lawson and C.-T. Yuen, *Eur.J Biochem.*, 2000, **267**, 1795.
9. B. K. Brandley, S. J. Swiedler and P. W. Robbins, *Cell*, 1990, **63**, 861.
10. T. Feizi, *Curr.Opin.Struct.Biol.*, 1991, **1**, 766.
11. M. P. Bevilacqua and R. M. Nelson, *J.Clin.Invest.*, 1993, **91**, 379.
12. T. Feizi, *Curr.Opin.Struct.Biol.*, 1993, **3**, 701.
13. C.-T.Yuen, A. M. Lawson, W. Chai, M. Larkin, M. S. Stoll, A. C. Stuart, F. X. Sullivan, T. J. Ahern and T. Feizi, *Biochemistry*, 1992, **31**, 9126.
14. W. Chai, T. Feizi, C.-T.Yuen and A. M. Lawson, *Glycobiology*, 1997, **7**, 861.

15. H. Kogelberg, T. A. Frenkiel, S. W. Homans, A. Lubineau and T. Feizi, *Biochemistry*, 1996, **35**, 1954.

16. C. Leteux, W. Chai, K. Nagai, A. M. Lawson, and T. Feizi, *J.Biol.Chem.*, 2001, **276**, 12539.

17. T. Osanai, T. Feizi, W. Chai, A. M. Lawson, M. L. Gustavsson, K. Sudo, M. Araki, K. Araki and C.-T. Yuen, *Biochem.Biophys.Res.Commun.*, 1996, **218**, 610.

18. P. Maly, A. D. Thall, B. Petryniak, C. E. Rogers, P. L. Smith, R. M. Marks, R. J. Kelly, K. M. Gersten, G. Cheng, T. L. Saunders, S. A. Camper, R. T. Camphausen, F. X. Sullivan, Y. Isogai, O. Hindsgaul, U. H. von Andrian and J. B. Lowe, *Cell*, 1996, **86**, 643.

19. C.-T. Yuen, W. Chai, R. W. Loveless, A. M. Lawson, R. U. Margolis and T. Feizi, *J.Biol.Chem.*, 1997, **272**, 8924.

20. W. Chai, C.-T. Yuen, H. Kogelberg, R. A. Carruthers, R. U. Margolis, T. Feizi and A. M. Lawson, *Eur.J.Biochem.*, 1999, **263**, 879.

21. J. Finne, T. Krusius, R. K. Margolis and R. U. Margolis, *J.Biol.Chem.*, 1979, **254**, 10295.

22. M. Larkin, T. J. Ahern, M. S. Stoll, M. Shaffer, D. Sako, J. O'Brien, A. M. Lawson, R. A. Childs, K. M. Barone, P. R. Langer-Safer, A. Hasegawa, M. Kiso, G. R. Larsen and T. Feizi, *J.Biol.Chem.*, 1992, **267**, 13661.

23. C. Galustian, R. A. Childs, C.-T. Yuen, A. Hasegawa, M. Kiso, A. Lubineau, G. Shaw and T. Feizi, *Biochemistry*, 1997, **36**, 5260.

24. C. Galustian, A. Lubineau, C. le Narvor, M. Kiso, G. Brown and T. Feizi, *J.Biol.Chem.*, 1999, **274**, 18213.

25. R. W. Loveless, C.-T. Yuen, H. Tsuiji, T. Irimura and T. Feizi, *Glycobiology*, 1998, **8**, 1237.

26. C. Leteux, W. Chai, R. W. Loveless, C.-T. Yuen, L. Uhlin-Hansen, Y. Combarnous, M. Jankovic, S. C. Maric, Z. Misulovin, M. C. Nussenzweig and T. Feizi, *J.Exp.Med.*, 2000 , **191**, 1117.

27. D. Fiete, M. C. Beranek and J. U. Baenziger, *Proc.Natl.Acad.Sci.U.S.A*, 1998, **95**, 2089.

28. Y. Liu, A. J. Chirino, Z. Misulovin, C. Leteux, T. Feizi, M. C. Nussenzweig and P. J. Bjorkman, *J Exp.Med.* , 2000, **191**, 1105.

29. I. J. Rosenstein, M. S. Stoll, T. Mizuochi, R. A. Childs, E. F. Hounsell and T. Feizi, *Lancet*, 1988, **ii**, 1327.

30. R. A. Childs, K. Drickamer, T. Kawasaki, S. Thiel, T. Mizuochi and T. Feizi, *Biochem.J.*, 1989, **262**, 131.

31. T. Mizuochi, R. W. Loveless, A. M. Lawson, W. Chai, P. J. Lachmann, R. A. Childs, S. Thiel and T. Feizi, *J.Biol.Chem.*, 1989, **264**, 13834.

32. M. Larkin, R. A. Childs, T. J. Matthews, S. Thiel, T. Mizuochi, A. M. Lawson, J. S. Savill, C. Haslett, R. Diaz and T. Feizi, *AIDS*, 1989, **3**, 793.

33. T. Feizi, *Immunol.Rev.*, 2000, **173**, 79.

34. C. Galustian, R. A. Childs, M. Stoll, H. Ishida, M. Kiso and T. Feizi, *Proceedings of the XVI International Symposium of Glycoconjugates, the Hague, August 2001*, eds. J. G. M. Bolchscher, I. van Die, J. P. Kammerling, G. A. Veldnik, J. F. G. Vleigenthart, Utreht. University, 2001, Abstract C2.27, p. 25.

Subject Index